# Cambridge Elements ≡

Elements in Decision Theory and Philosophy
edited by
Martin Peterson
*Texas A&M University*

# DUTCH BOOK ARGUMENTS

Richard Pettigrew
*University of Bristol*

CAMBRIDGE
UNIVERSITY PRESS

# CAMBRIDGE
## UNIVERSITY PRESS

University Printing House, Cambridge CB2 8BS, United Kingdom

One Liberty Plaza, 20th Floor, New York, NY 10006, USA

477 Williamstown Road, Port Melbourne, VIC 3207, Australia

314–321, 3rd Floor, Plot 3, Splendor Forum, Jasola District Centre, New Delhi – 110025, India

79 Anson Road, #06–04/06, Singapore 079906

Cambridge University Press is part of the University of Cambridge.

It furthers the University's mission by disseminating knowledge in the pursuit of education, learning, and research at the highest international levels of excellence.

www.cambridge.org
Information on this title: www.cambridge.org/9781108713443
DOI: 10.1017/9781108581813

© Richard Pettigrew 2020

First published 2020

*A catalogue record for this publication is available from the British Library.*

ISBN 978-1-108-71344-3 Paperback
ISSN 2517-4827 (online)
ISSN 2517-4819 (print)

# Dutch Book Arguments

Elements in Decision Theory and Philosophy

DOI: 10.1017/9781108581813
First published online: July 2020

Richard Pettigrew
*University of Bristol*

**Author for correspondence:** Richard Pettigrew
richard.pettigrew@bristol.ac.uk

**Abstract:** Our beliefs come in degrees. I'm 70 per cent confident it will rain tomorrow, and 0.001 per cent sure my lottery ticket will win. What's more, we think these degrees of belief should abide by certain principles if they are to be rational. For instance, you shouldn't believe that a person's taller than 6 ft more strongly than you believe that they're taller than 5 ft, since the former entails the latter. In Dutch Book arguments, we try to establish the principles of rationality for degrees of belief by appealing to their role in guiding decisions. In particular, we show that degrees of belief that don't satisfy the principles will always guide action in some way that is bad or undesirable. In this Element, we present Dutch Book arguments for the principles of Probabilism, Conditionalisation, and the Reflection Principle, among others, and we formulate and consider the most serious objections to them.

**Keywords:** Dutch Book arguments, Bayesian epistemology, Bayesianism, Probabilism, probability

ISBNs: 9781108713443 (PB), 9781108581813 (OC)
ISSNs: 2517-4827 (online), 2517-4819 (print)

# Contents

# 1 Overview

Our beliefs come in degrees. I believe some things more strongly than I believe others. I believe very strongly that the average global temperature at sea level will continue to rise during the coming century; I believe slightly less strongly that the European Union will still exist in 2029; and I believe much less strongly that Cardiff is east of Edinburgh. My credence in something is a measure of the strength of my belief in it; it represents my level of confidence in it. These are the states of mind we report when we say things like 'I'm 20 per cent confident I switched off the gas' or 'I'm 99.9 per cent sure that it is raining outside'. There are laws that govern these credences. For instance, I shouldn't be more confident that sea levels will rise by over 2 metres in the next 100 years than I am that they'll rise by over 1 metre, since the latter is true if the former is. This Element is about a particular way we might try to establish these laws of credence, namely, the Dutch Book arguments.[1]

Dutch Book arguments originate in two seminal papers by mathematicians from the first half of the twentieth century: Frank P. Ramsey hints at one version in his 'Truth and Probability', while Bruno de Finetti spells out another in much greater detail and using the mathematical techniques that we'll deploy in this Element in his 'Foresight: Its Logical Laws, Its Subjective Sources' (Ramsey, 1926 [1931]; de Finetti, 1937 [1980]). In each such argument, we assume that your credences are related in some way to your decisions – those credences will or should lead you to make certain choices faced with certain decisions. And we show that credences that violate the law we wish to establish are related in that way to decisions that are guaranteed to serve our ends badly. On that basis, we conclude that such credences are irrational, and we take ourselves to have established the law. So, Dutch Book arguments evaluate the rationality of credences by looking at the quality of the choices that they do or should lead us to make. Typically, but not always, the decisions in question are betting decisions – whether or not to pay a particular price to enter into a bet on a proposition that pays out a certain amount if the proposition is true and pays out nothing if it is false. And, in the original Dutch Book argument for the law of

---

[1] For briefer overviews of these arguments, see Hájek (2008) and Vineberg (2016). There is much speculation but little known fact about the origin of the name. Some say it alludes to a practice amongst Dutch insurance companies in the nineteenth century, whereby they arranged the premiums and payouts of their schemes in such a way that they were guaranteed to gain. Others say its origins lie in horse racing terminology, though the term 'dutching' that is used there describes something different. We retain the term here because it has become so deeply embedded in the literature. But I recognise that it isn't ideal, partly because it doesn't give much information to the uninitiated about what these arguments involve, and partly because some feel that it is disparaging to Dutch people. Alternatives that avoid those problems might be *sure loss arguments* or simply *betting arguments*.

Probabilism that Ramsey gestures towards and de Finetti formulates precisely, we show that credences that violate that law are related to a series of betting decisions that, taken together, are guaranteed to lose you money – that is, the total price you pay for the bets exceeds the highest total pay-off you might receive from them. Such choices clearly serve your ends poorly, and thus we conclude that the credences that are related to them are irrational.

Before we begin, let me describe the structure of this Element. Dutch Book arguments are formulated in many different ways throughout the philosophical literature. To home in on the best possible version, we begin, in Section 2, with one of the most straightforward and widely used formulations of the Dutch Book arguments for a variety of laws that are thought to govern our credences: arguments for some of the central tenets of Bayesian epistemology, such as Normalisation and Finite Additivity, which are together known as Probabilism; and arguments for some additional putative laws, which are accepted by some Bayesians and not by others, namely, Countable Additivity, Regularity, and the Principal Principle. Then, in Section 3, we subject these standard formulations to rigorous stress-testing, and we gradually amend them little by little so that they can withstand various objections. What we are left with is still recognisably the orthodox Dutch Book arguments for nearly all of these laws, but now in a more resilient formulation. However, we show that the argument for Countable Additivity cannot work and we pursue it no further (Section 3.7). Those already familiar with Dutch Book arguments might wish to skip forward to Section 3.9, where we spell out our final formulations of Probabilism, Regularity, and the Principal Principle and present our final formulations of the standard Dutch Book arguments in their favour.

Next, in Section 4, we set out the Dutch Strategy argument for another of the central tenets of Bayesian epistemology, namely, Conditionalisation. We note that this argument can only establish a version of that law that applies to the updating rules we are disposed to follow, and not to our actual updating behaviour. And we respond to a standard objection, which argues that, if the Dutch Strategy argument for Conditionalisation works, then so does the Dutch Strategy argument for the Reflection Principle, and then adds that the Reflection Principle is false.

In Sections 5 and 6, we consider three objections to Dutch Book arguments that cannot be addressed by making small adjustments. In Section 5, we ask whether our credences really require of us what the Dutch Book arguments assume they do. We ask what the correct version of expected utility theory is for individuals whose credences violate some of the norms of Bayesian epistemology; and then we ask how the Dutch Book argument fares if we use a non-expected utility theory to determine the price an individual should pay

for a bet. In Section 6, we ask whether it is really irrational to have credences that are related to prices for bets that are guaranteed to be excessive in the way the Dutch Book argument identifies. In response to the first and third of these three objections, we must completely redesign those arguments, replacing them with ones that share a general approach but few specific details; I can find no adequate response to the second, and I leave this as an open question.

In Section 7, we ask what happens to the Dutch Book arguments if we change certain features of the basic framework in which we've been working. First, we ask how Dutch Book arguments fare when we consider credences in self-locating propositions, such as *It is Monday*. Second, we lift the assumption that the background logic is classical and explore Dutch Book arguments for non-classical logics, such as strong Kleene logic, the Logic of Paradox, and supervaluationist semantics. And, third, we lift the assumption that an agent's credal state can be represented by a single assignment of numerical values to the propositions she considers, and we explore how Dutch Book arguments work when we represent individuals using imprecise credences.

In Section 8, we present the mathematical results that underpin these arguments.

## 2 Introducing Dutch Book Arguments

Finella is a climate scientist. She specialises in predicting how much the average temperature of Earth at sea level will change in the future. Today, she's thinking about the change over the coming 100 years. She divides the possibilities into three categories. A change of at most 0°C counts as *low*, a change of between 0°C and 1°C counts as *medium*, and a change of at least 1°C counts as *high*. She's 30 per cent confident that the change will be low and 40 per cent confident it will be medium. But she's also 90 per cent confident it won't be high. As a result, there seems to be something wrong with her credences. After all, the temperature change is not high just in case it is either low or medium. So, if she is rational, we would surely expect her credence that the change won't be high to be the sum of her credence that it will be low and her credence that it will be medium. But it isn't. (See Figure 2.1 below for an illustration of Finella's credences along with those of her colleagues.)

Norma is Finella's colleague. She's less than 90 per cent confident that the change in temperature will be either low, medium, or high. There seems to be something wrong with her credence too. After all, it's necessarily true that the change will be low, medium, or high – those three categories cover all the possibilities. And, if she's rational, we expect her to be 100 per cent confident in something that is necessarily true. But she isn't.

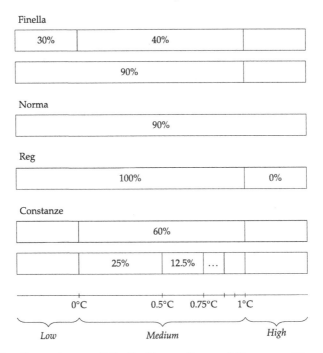

**Figure 2.1** The credences of Finella, Norma, Reg, and Constanze. The credences of Pritpal are more difficult to illustrate.

Reg is another meteorologist. He's 0 per cent confident that the temperature change will be high and 100 per cent confident it won't be. That is, he concentrates all his credence on the possibility of a low or a medium change. But he has no evidence that conclusively rules out a high change. Again, there seems to be something wrong with Reg's credences. If he were rational, we'd expect him to assign at least some credence to everything that is compatible with his evidence. But he doesn't.

Constanze is looking at a rather more fine-grained set of possibilities than her three colleagues. She is considering not just the three possibilities – low, medium, and high – but also an infinite series of possibilities within the medium range of 0°C to 1°C. She's considering the range more than 0°C to less than 0.5°C, at least 0.5°C to less than 0.75°C, at least 0.75°C to less than 0.875°C, and so on. These ranges don't overlap, and together they cover the entire medium range. Constanze's credence that the change will be medium is 60 per cent. Her credence that it will lie in the range from 0°C to 0.5°C is 25 per cent, from 0.5°C to 0.75°C is 12.5 per cent, from 0.75°C to 0.875°C is 6.25 per cent, and so on. The infinite sum of her degrees in the more fine-grained possibilities is thus 50 per cent. If she were rational, we might expect that sum to be equal to

her credence in the full range that these possibilities cover, namely, the medium range from 0°C to 1°C. But it isn't.

Pritpal is setting his credence that Storm Saoirse will make landfall in the UK by midnight. To do this, he investigates the objective chance for that event.[2] His investigation isn't conclusive. His evidence doesn't tell between a chance of 50 per cent, which is suggested by one of his models, and a chance of 70 per cent, suggested by another. In the light of this limitation, he decides to set his conditional credence for that event given that the chances are one way, and his conditional credence given that they are the other way. His credence in landfall before midnight, given that the chance of landfall before midnight is 50 per cent, is 40 per cent. Like his colleagues, it seems that Pritpal is irrational. If he were rational, we might expect his credence in landfall before midnight given that the chance of that event is 50 per cent to be 50 per cent. But it isn't.

In each of these five cases, many people feel that the individuals' credences are irrational. But why? What is irrational about them? It is just such questions that Dutch Book arguments are intended to answer. They are intended to establish that certain ways of thinking are irrational – they are intended to establish certain laws of thought.

Now establishing the laws that govern our credences is an important task. After all, we reason with them all the time. My doctor reasons with her credences in various different diagnoses as she weighs up the evidence provided by my blood tests and my ultrasound scan. Finella, Norma, and their colleagues reason with their credences when they take the evidence from current computer models, together with data from the near and far past climate, and come up with their best estimate of the change in global mean surface temperature over the coming century. And, in much more prosaic cases, I use my credences when I try to figure out my opponent's hand in poker, or my best estimate for the weight of the prize marrow at the local county fair. What's more, our credences aren't just used in reasoning that results in further credences or estimates; they are also used in reasoning that issues in action. A government minister uses their credences to reason about whether or not to build a flood defence in a particular area. My doctor uses her credences to decide whether or not to order further tests. And I use my credences about the location of my house keys to decide where to search for them first.

---

[2] Just as an individual's credence in a proposition measures how likely they take that proposition to be, so the objective chance of a proposition measures how likely the world makes it. These are the facts about the world we report when we say things like 'This sample of uranium has a 10 per cent chance of decaying in the next hour' or 'If you are a woman with a mutation in the PALB2 gene, your chance of developing breast cancer by the time you are 70 years old is 33 per cent'.

In this Element, we'll present a wide range of Dutch Book arguments for an equally wide range of laws of credence. We'll start here by sketching very crude versions of the Dutch Book arguments for five such laws: Normalisation, Finite and Countable Additivity, Regularity, and the Principal Principle. Then we'll present a series of objections to them, amend them to avoid those objections, and emerge at the end of the next section with much improved arguments.

## 2.1 The Laws of Credence

Normalisation is the law that Norma violates:

> **Normalisation** Your credence in a necessarily true proposition should be the maximal possible credence (i.e. 100 per cent).

Note that this assumes that there is a maximal credence you can have in a proposition; it assumes there is an upper bound to the strength of the beliefs you can hold. We might call this maximal credence *certainty*.

Finite Additivity is the law that Finella violates:

> **Finite Additivity** If $A$ and $B$ are two propositions that can't both be true, your credence in their disjunction, $A \vee B$, should be the sum of your credence in $A$ and your credence in $B$.

Countable Additivity is the law that Constanze violates:

> **Countable Additivity** If $A_1, A_2, \ldots$ are infinitely many propositions no two of which can both be true, your credence in their disjunction, $A_1 \vee A_2 \vee \ldots$, should be the infinite sum of your credences in $A_1, A_2, \ldots$.

Regularity is the law that Reg violates:

> **Regularity** If $A$ is a proposition that is possibly true, then your credence in $A$ should be positive.

The Principal Principle is the law that Pritpal violates:

> **Principal Principle** If you have a positive credence that the objective chance of $A$ is $x$ per cent, then your credence in $A$ conditional on the objective chance of $A$ being $x$ per cent should be $x$ per cent.

Together, Finite Additivity and Normalisation make up the law of Probabilism, which says that an individual's credences should obey the axioms of the probability calculus.

## 2.2 Representing Credences

Here and throughout this Element, we will translate percentage levels of confidence into credences that are at least 0 and at most 1, so that 0 per cent confidence translates to a credence of 0, 100 per cent to a credence of 1, 30 per cent to 0.3, and so on. Thus, Norma's credence in the disjunction *Low* ∨ *Medium* ∨ *High* is 0.9, Reg's credence in *Low* is 0 and his credence in the negation of *Low* is 1, and so on.

## 2.3 Betting Prices

The Dutch Book arguments for Normalisation, Finite and Countable Additivity, Regularity, and the Principal Principle all share the same first premise. It describes a connection between your credences and the bets you'll enter into. Consider a bet on a proposition that pays out £100 if the proposition is true, and £0 if it is false. We'll call this a *£100 bet* on that proposition. Then minimal credence (i.e. 0 per cent or 0) in that proposition will lead you to buy or sell a £100 bet for £0, maximal credence (i.e. 100 per cent or 1) will lead you to buy or sell it for £100, and for any credence in between, the price you'll buy or sell it for increases linearly with the credence. So, if you have a credence of 20 per cent or 0.2 in a proposition, you'll buy or sell a £100 bet on that proposition for £20, and if you have a credence of 75.6 per cent or 0.756, you'll buy or sell for £75.60.

In general, if $S$ is a real number – positive, negative, or zero – a $£S$ bet on a proposition is one that pays $£S$ if the proposition is true and £0 if it is false. And, if you have credence $p$ in a proposition, then you'll buy or sell a $£S$ bet on that proposition for $£pS$. Following is some terminology that appears in the literature on Dutch Book arguments:

- The *stake* of a $£S$ bet is $£S$.
- Your *betting quotient* for a proposition is the real number $q$ such that, for any stake $S$, you'll pay $£qS$ for a $£S$ bet on that proposition. (Note that it is a significant assumption that there is any such number.)

So the assumption we have introduced here might be restated as follows: your credence in a proposition always matches your betting quotient for it. Both Ramsey and de Finetti held that, while this might not be strictly true in all cases, it is at least a useful idealisation. We will discuss it at length at the beginning of Section 3.

Thus, if your credence in *Low* is 40 per cent or 0.4, you'll pay 40p for a £1 bet on *Low*. The stake of this bet of £1.

- The net gain of this bet if *Low* is true will be its payout in that situation (£1) minus its price (40p). That is, it will be 60p.
- The net gain if *Low* is false will be its payout in this situation (£0) minus its price (40p). That is, it will be −40p.

Similarly, if your credence in *Low* is 60 per cent or 0.6, you'll pay −60p for a −£1 bet on *Low*. That is, you'll sell a £1 bet on *Low* for 60p. The stake of this bet is −£1.

- The net gain of this bet if *Low* is true will be its payout (−£1) minus its price (−60p). That is, it will be −40p.
- The net gain if *Low* is false will be its payout (£0) minus its price (−60p). That is, it will be 60p.

In general, the net gain of a £$S$ bet on $X$ at price £$pS$ is given in the following table:

|  | Net gain |
| --- | --- |
| $X$ | $(1-p)S$ |
| $\overline{X}$ | $-pS$ |

where $\overline{X}$ is the negation of $X$.

## 2.4 Norma and the Law of Normalisation

Now, let's apply this premise to Norma's credence of 0.9 in the disjunction *Low* ∨ *Medium* ∨ *High* – a credence that violates Normalisation. It says that she will sell a £100 bet on the disjunction for £90. Put differently, she will accept a bet with the following payouts (in pounds):

$$Bet\ 1 = \begin{cases} 90 - 100 = -10 & \text{if the disjunction is true} \\ 90 & \text{if the disjunction is false} \end{cases}$$

Thus, if the disjunction is true, Norma will receive £90 and pay out £100. So her net loss will be £10. If, on the other hand, it is false, Norma will receive £90 and pay out £0. So her net gain will be £90. Unfortunately for Norma, however, the disjunction is necessarily true. So she is guaranteed to make a net loss. However the world turns out, she'll lose £10. That observation is the argument's second premise.

Now, it was her credence in *Low* ∨ *Medium* ∨ *High* that led her to enter into that bet at that price. And thus, the Dutch Book argument for Normalisation concludes, her credences must be irrational. It is irrational, the argument's third

premise contends, to have credences that lead you to enter into bets at prices that, taken together, are guaranteed to lose you money. A series of bets at such prices are known as *Dutch Books* – they give the species of argument its name.

De Finetti said that credences are *incoherent* if they will lead you to enter into a Dutch Book (de Finetti, 1937 [1980], 103). Thus, we have just shown that Norma's credence in the disjunction *Low* ∨ *Medium* ∨ *High* is incoherent. For de Finetti, incoherence is a form of inconsistency.

Of course, the present argument only provides a Dutch Book against Norma's particular violation of Normalisation. But it is easy to see how to adapt it to generate a Dutch Book against any other violation. In de Finetti's terminology, we show that any violation of Normalisation is incoherent. If your credence $p$ in a necessary truth $\top$ is less than maximal, then, for any real number $S$, you'll sell a £$S$ bet on $\top$ for £$pS$, which is below £$S$. But this will lose you money if $\top$ is true, which it is guaranteed to be, and you'll only gain money if $\top$ is false, which it is guaranteed not to be. In table form:[3]

|  | Net gain |
|---|---|
| $\top$ | $-(1-p)S$ |
| $\bot$ | $pS$ |

If $p < 1$, then let the stake be positive, that is, $S > 0$. That ensures that your net gain is necessarily negative, since $-S(1-p) < 0$.

## 2.5 Finella and the Law of Finite Additivity

Let's turn next to Finella – her credences violate Finite Additivity:

| *Low* | *Medium* | *Low* ∨ *Medium* |
|---|---|---|
| 0.3 | 0.4 | 0.9 |

The first premise of the Dutch Book argument tells us that she'll accept the following bets:

- Finella's credence of 0.3 in *Low* will lead her to sell a £100 bet on *Low* for £30 (Bet 1).
- Finella's credence of 0.4 in *Medium* will lead her to sell a £100 bet on *Medium* for £40 (Bet 2).

---

[3] Here and throughout, $\top$ is a necessarily true proposition, while $\bot$ is a necessarily false one.

- Finella's credence of 0.9 in *Low* ∨ *Medium* will lead her to buy a £100 bet on *Low* ∨ *Medium* for £90 (Bet 3).

Let's see what her net gain or loss will be, having accepted this book of bets:

|         | Bet 1 | Bet 2 | Bet 3 | Net gain |
|---------|-------|-------|-------|----------|
| *Low*    | −70   | 40    | 10    | −20      |
| *Medium* | 30    | −60   | 10    | −20      |
| *High*   | 30    | 40    | −90   | −20      |

That is, however the world turns out, however the temperature changes over time, Finella will lose £20 from having entered into these bets at the prices in question. But, again, it was her credences that made her do it! And thus, again, the Dutch Book argument concludes that her credences must be irrational.

Of course, the present argument only provides a Dutch Book against Finella's particular violation of Finite Additivity. But it is easy to see how to adapt it to generate a Dutch Book against any other violation. That is, we can show that any violation of Finite Additivity is incoherent in de Finetti's terminology. Suppose $r$ is your credence in an exclusive disjunction $A \vee B$, $p$ is your credence in $A$, and $q$ is your credence in $B$. Then you'll pay $£pS$ for a $£S$ bet on $A$ (Bet 1), $£qS$ for a $£S$ bet on $B$ (Bet 2), and $-£rS$ for a $-£S$ bet on $A \vee B$ (Bet 3). Then your total net gains will be:

|                  | Bet 1      | Bet 2      | Bet 3       | Net gain          |
|------------------|------------|------------|-------------|-------------------|
| $A\overline{B}$  | $(1-p)S$   | $-qS$      | $-(1-r)S$   | $(r-(p+q))S$      |
| $\overline{A}B$  | $-pS$      | $(1-q)S$   | $-(1-r)S$   | $(r-(p+q))S$      |
| $\overline{AB}$  | $-pS$      | $-qS$      | $rS$        | $(r-(p+q))S$      |

Thus, if $p+q < r$, then let $S < 0$, and if $p+q > r$, then let $S > 0$. That ensures that your total net gain is necessarily negative.

## 2.6 Constanze and the Law of Countable Additivity

The Dutch Book argument against Constanze is very similar to the argument against Finella (Williamson, 1999). Recall her credences:

Constanze will buy a £100 bet on the range 0°C to 1°C for £60, she'll sell a £100 bet on the range 0°C to 0.5°C for £25, a £100 bet on the range 0.5°C to 0.75°C for £12.50, a £100 bet on 0.75°C to 0.875°C for £6.25, and so on. The

| 0°C–1°C | 0°C–0.5°C | 0.5°C–0.75°C | 0.75°C–0.875°C | ... |
|---|---|---|---|---|
| 0.6 | 0.25 | 0.125 | 0.0625 | ... |

total she'll receive from selling that infinite series of bets (£50) will be less than the amount she pays out for the bet that she buys (£60). So she will be down £10 before the bets pay out. When the bets pay out, if the temperature increase is in the medium range, she'll receive £100 from the bet she bought on that possibility, and she'll pay out £100 for the single bet on whichever of the more fine-grained possibilities within that range is true. These will cancel and she'll end up £10 down overall. On the other hand, if the temperature increase is not in the medium range, she'll receive nothing from the bet she bought and pay out nothing from the bets she sold. So, again, she'll end up £10 down overall.

Again, this generalises so that we can generate a Dutch Book against any violation of Countable Additivity. Suppose $A_1, A_2, \ldots$ is an infinite list of disjoint propositions – that is, $A_i$ & $A_j$ is necessarily false for all $i \neq j$. And let $A$ be their infinite disjunction – that is, $A$ is true exactly when one of $A_1, A_2, \ldots$ is true. Suppose $p_i$ is your credence in $A_i$ and $p$ is your credence in $A$. Then you'll sell a £$S$ bet on $A$ for £$pS$ (Bet 0), and, for each $i$, you'll buy a £$S$ bet on $A_i$ for £$p_iS$ (Bet $i$). Your total net gains will be as follows:

|  | Bet 0 | Bet 1 | Bet 2 | Bet 3 | ... | Net gain |
|---|---|---|---|---|---|---|
| $\overline{A}$ | $pS$ | $-p_1S$ | $-p_2S$ | $-p_3S$ | ... | $(p - \sum_{i=1}^\infty p_i)S$ |
| $A_1$ | $-(1-p)S$ | $(1-p_1)S$ | $-p_2S$ | $-p_3S$ | ... | $(p - \sum_{i=1}^\infty p_i)S$ |
| $A_2$ | $-(1-p)S$ | $-p_1S$ | $(1-p_2)S$ | $-p_3S$ | ... | $(p - \sum_{i=1}^\infty p_i)S$ |
| $A_3$ | $-(1-p)S$ | $-p_1S$ | $-p_2S$ | $(1-p_3)S$ | ... | $(p - \sum_{i=1}^\infty p_i)S$ |
| ⋮ | ⋮ | ⋮ | ⋮ | ⋮ | ... | ⋮ |

Now, if $\sum_{i=1}^\infty p_i < p$, then let $S < 0$, and if $\sum_{i=1}^\infty p_i > p$, then let $S > 0$. That ensures that your total net gain is necessarily negative.

## 2.7 Reg and the Law of Regularity

The Dutch Book argument against Reg is very similar to the argument against Norma. *High* is possible, but Reg assigns credence 0 to it. So he'll sell a £100 bet on *High* for £0. Now, this bet cannot gain him money, but it can lose him money – it loses him money if *High* is true, and by hypothesis that's a possibility. Again, we conclude that Reg's credence is irrational, since it leads him to enter into a bet with a possibility of loss but no possibility of gain – we call such a bet or series of bets an *almost-Dutch Book*.

Again, this generalises so that we can generate an almost-Dutch Book against any violation of Regularity. Suppose $A$ is possible, but you assign it credence 0. Then you'll sell a £$S$ bet on $A$ for £0. Your net gain:

|   | Net gain |
|---|---|
| $A$ | $-S$ |
| $\overline{A}$ | $0$ |

So let $S > 0$ to ensure that your net gain is necessarily not positive and possibly negative.

## 2.8 Pritpal and the Principal Principle

The Dutch Book argument against Pritpal is a little different – and a whole lot more involved. Instead of giving the argument against his particular credences and then generalising, as we have with Finella, Norma, and so on, it will be more straightforward to move straight to the general version. Those interested only in the Dutch Book arguments for the other norms can skip past this section without loss.

Let $C_r^A$ be the proposition that the chance of proposition $A$ is $r$. Then the Principal Principle says that if your credence in $C_r^A$ is positive, then your conditional credence in $A$ given $C_r^A$ should be $r$. So, suppose $p$ is your credence in the conjunction $A$ & $C_r^A$, while $q$ is your credence in $C_r^A$. Then, if your credence in $C_r^A$ is positive, we say that your conditional credence in $A$ given $C_r^A$ is just the ratio of $A$ & $C_r^A$ to $C_r^A$. Thus, if $q > 0$, you satisfy the Principal Principle just in case $\frac{p}{q} = r$. Now, if $q > 0$, you will accept the following bets:

- Sell a £$S$ bet on $A$ & $C_r^A$ for £$pS$ (Bet 1);
- Buy a £$\frac{r+\frac{p}{q}}{2}S$ bet on $C_r^A$ for £$q\frac{r+\frac{p}{q}}{2}S$ (Bet 2).

Taken together, these bets have the payouts indicated in Table 1.
Now, suppose $C_r^A$ is true. That is, the chance of $A$ is indeed $r$. Then we can assume that:

- the chance of $A$ & $C_r^A$ is also $r$,
- the chance of $\overline{A}$ & $C_r^A$ is $1 - r$, and
- the chance of $\overline{C_r^A}$ is 0.[4]

---

[4] In fact, this is not an entirely innocent assumption, as we will see when we consider a more general version of the Principal Principle below. It amounts to assuming, in David Lewis' terminology, that the chances are not *self-undermining* (Lewis, 1980).

**Table 1**

| | Bet 1 | Bet 2 | Net gain |
|---|---|---|---|
| $A \,\&\, C_r^A$ | $-(1-p)S$ | $(1-q)\frac{r+\frac{\ell}{q}}{2}S$ | $\left(-(1-p)+(1-q)\frac{r+\frac{\ell}{q}}{2}\right)S$ |
| $\overline{A} \,\&\, C_r^A$ | $pS$ | $(1-q)\frac{r+\frac{\ell}{q}}{2}S$ | $\left(p+(1-q)\frac{r+\frac{\ell}{q}}{2}\right)S$ |
| $\overline{C_r^A}$ | $pS$ | $-q\frac{r+\frac{\ell}{q}}{2}S$ | $p-q\frac{r+\frac{\ell}{q}}{2}S$ |

Thus, the objective chances expect Bets 1 and 2, taken together, to have the following monetary value:

$$r\left(-(1-p)+(1-q)\frac{r+\frac{\ell}{q}}{2}\right)S+(1-r)\left(p+(1-q)\frac{r+\frac{\ell}{q}}{2}\right)S=(1+q)\frac{\frac{\ell}{q}-r}{2}S$$

Next, suppose $C_r^A$ is false. Then, whatever the chances are:

- the chance of $A \,\&\, C_r^A$ is 0,
- the chance of $\overline{A} \,\&\, C_r^A$ is 0, and
- the chance of $C_r^A$ is 1.

So the expected monetary values of the bets is just their monetary value when $\overline{C_r^A}$ is true, which is

$$p-q\frac{r+\frac{\ell}{q}}{2}S=q\frac{\frac{\ell}{q}-r}{2}S$$

Now, if you violate the Principal Principle, $\frac{\ell}{q} \neq r$. If $\frac{\ell}{q} < r$, let $S > 0$, and if $\frac{\ell}{q} > r$, then let $S < 0$. Then, whatever the chances are, they expect Bets 1, 2, and 3, taken together, to lose you money – we call such a series of bets an *expectedly-Dutch Book*.

Notice that expectedly-Dutch Book arguments are rather different from other Dutch Book arguments. In the other arguments, we show that someone who violates the law in question will accept bets that, taken together, lose them money for sure (or at least fail to gain them money for sure and create the possibility of losing them money). This is not the case for the argument we have just presented. A quick calculation shows that, if the stake $S$ is positive, and if $A$ is false and the chance of $A$ is $r$ (i.e. $\overline{A} \,\&\, C_r^A$), then your net gain is non-negative. But what is guaranteed is that the chances will expect you to lose money. That is, whether the chance of $A$ is $r$ or not, you will be expected to lose money.

### 3 Rewriting the Dutch Book

We have now sketched five very crude Dutch Book arguments, one each for Normalisation, Finite and Countable Additivity, Regularity, and the Principal Principle. The arguments for Normalisation and Finite Additivity together give a Dutch Book argument for Probabilism, which is the conjunction of those two laws. In this section, we'll fix up some basic flaws in these arguments as I have formulated them. This will leave us with reasonably robust versions that we can then subject to more searching critiques in the sections that follow.

## 3.1 Normative and Descriptive Readings

As we have presented it here, in its first premise, a Dutch Book argument assumes that, for instance, someone with credence 0.7 in a proposition *will* buy or sell a £100 bet on that proposition for £70. But will they? It seems that only two things could guarantee this.

First, it might be true by definition. We might say that what it means to say that you have credence 0.7 in a proposition is that you will accept such a bet. And in general, we might say that what it means to have a certain credal attitude to a proposition is to be disposed to behave in certain ways when faced with decisions whose outcomes depend on the truth or falsity of that proposition. But this can't be right. As David Christensen (1996) notes, even if credences are defined in terms of their connections to other aspects of our mental state and our behaviour, as the functionalist would have us believe, they will not be defined in terms of their connections to action alone. Our credences are connected also to our emotions, for instance, as well as to one another. When I have a high credence that danger awaits around the next corner, it often causes me to feel fear; when I have low confidence my partner loves me, I usually feel sorrow; when I have high credence that I am in Edinburgh, I usually have even higher confidence that I'm in Scotland; and so on. But note that these connections are not exceptionless. Indeed, Dutch Book arguments are concerned with exactly those cases in which the connections between one credence and another are broken – a case in which I have a high credence that I am in Edinburgh but a low credence that I am in Scotland, for instance. So, the functionalist does not say that you have a credence iff you have a mental state that invariably causes such-and-such other states and actions; they say that you have a credence iff you have a mental state that enters into sufficiently many of these connections sufficiently often. So, for the functionalist, it is quite possible to have a credence in a proposition and yet not exhibit the behaviour that the Dutch Book argument assumes you'll exhibit. So this first route to justifying the first premise of the Dutch Book argument is blocked.

Second, it could be that, while it is not true by definition that having a credence will lead to certain behaviour, it is nonetheless always true. This could perhaps be because of some law of nature that guarantees this regularity. There are two things to say about this: First, it simply doesn't seem to be true. Some people have an aversion to betting, even when they think the price offered for the bet is fair or to their advantage; others might take the fact that the bet is offered as evidence that they are being tricked and might opt out on that basis (Ramsey, 1926 [1931], 172). Second, however, even if it were true, it isn't clear that it is irrational to have mental states that unfailingly lead you to make bad decisions. For instance, suppose I exhibit a strange pathology: the minute I fall in love, I wolf down a meal of mud, leaves, and earthworms. It's an exception-less regularity. But this does not show that it is irrational for me to fall in love. Rather, it is irrational for me to *respond* to falling in love in this way. So, even if it were true that having a credence in a proposition always leads me to accept the bets specified in the Dutch Book argument, it doesn't follow that having credences that lead me to accept a Dutch Book are irrational. It could be that it is my response to having those credences that is irrational.

This brings us to our reformulation of that first premise. Instead of saying that someone with credence 0.7 in a proposition *will* buy or sell a £100 bet on that proposition for £70, we say instead that they *should* buy or sell that bet at that price. And that is surely enough. It is surely irrational to have credences that *should* lead you to accept each of a series of bets that, when taken together, are guaranteed to lose you money. It is surely irrational to have credences that *rationally require* you to enter into a Dutch Book.

## 3.2 Sweetening the Pill

Having tweaked the first premise of the argument in this way, we face a different problem. Suppose you have credence 0.7 in a proposition $X$. Then our new version of the first premise of the Dutch Book argument says that you should pay £70 for a £100 bet on $X$. But let's suppose, moreover, that you have credence 0.3 in $\overline{X}$, the negation of $X$, just as Probabilism says you should. Then, in that situation, it's natural to think that you should choose in line with your expectations. And the expected monetary value of the bet in question is $£(0.7 \times (-70 + 100) + 0.3 \times (-70)) = £0$, which is the same as the expected monetary value of refusing the bet. So, you might reasonably be indifferent between paying that price for the bet and refusing it. And if that's the case, it simply isn't true that you *should* accept this bet. At the most, you *may* accept it. It is not true that you are rationally *required* to take the bet. At most, you are rationally *permitted*.

We might respond to this objection by noting that, while it might be more irrational to have credences that *rationally require* you to accept a Dutch Book, an almost-Dutch Book, or an expectedly-Dutch Book, it is still irrational to have credences that *rationally permit* you to accept each bet in such a set. However, this would lead to a weaker conclusion, since it imputes a weaker sort of irrationality to someone who violates a law of credence.

Another response is to replace each of the bets in the Dutch Book arguments with ones that are rationally required by the credences in question. Which are these? The following will be the first premise of our new Dutch Book arguments:

(DB1)  If you have credence $p$ in proposition $A$, then

  (i)   you are rationally permitted to pay $£x$ for a $£S$ bet on $A$ for any $x \leq pS$,
  (ii)  you are rationally required to pay $£x$ for a $£S$ bet on $A$ for any $x < pS$.

Now, take Finella, for instance. Here are her credences, which violate Finite Additivity:

| Low | Medium | Low ∨ Medium |
|-----|--------|--------------|
| 0.3 | 0.4    | 0.9          |

While she might be rationally permitted but not required to sell a £100 bet on *Low* for £30, she's rationally required to sell it for £35 (Bet 1'). Similarly, she's required to sell a £100 bet on *Medium* for £45 (Bet 2'). And she's required to buy a £100 bet on *Low* ∨ *Medium* for £85 (Bet 3'). Each of these is derived from the bets in the original Dutch Book arguments against Finella by raising the price of those she will sell by £5, and lowering the price of those she'll buy by £5. Each has a positive expected monetary value of £2. And, furthermore, taken together, the bets lose Finella money for sure. However the world turns out, she will suffer a total net loss of £5. And we can do the same for the Dutch Books against Norma, Constanze, Reginald, and Pritpal. In each case, there's always a small positive amount by which we can increase the price of bets the individual sells and decrease the price of bets they buy while ensuring that, taken together, the bets still lose the individual money for sure.

## 3.3 The Diminishing Marginal Value of Money

Part of the attraction of this new version of the first premise in our Dutch Book arguments is that, if your credence in $A$ is $p$ and your credence in its negation $\overline{A}$ is $1 - p$, as Probabilism requires, then the expected monetary value of paying £$x$ for a £$S$ bet on $X$ is £$(p(S - x) + (1 - p)(0 - x)) = £(pS - x)$, and this is equal to the expected monetary value of not paying for the bet (i.e. £0) if $x = pS$ and greater than that expected monetary value if $x < pS$. So, the first premise of our Dutch Book arguments is consistent with the demand to choose so as to maximise expected monetary value. But orthodox decision theory does not demand that you choose so as to maximise expected *monetary value*; it demands you choose so as to maximise expected *utility*. And, as Daniel Bernoulli and Gabriel Cramer observed early in the history of the subject, these two come apart because the utility of having a certain amount of money might not be a linear function of that money (Bernoulli, 1738 [1954]). That is, for many individuals, how much extra utility you gain by receiving £100 will depend on how much money you have already, so that an extra £100 might make an enormous difference to a person on a low income in the UK, while it would make almost no difference to the CEO of a major corporation in the USA.

Suppose, for instance, that Finella's utility is a logarithmic function of her monetary wealth. Thus, if her wealth is £$k$, then, measured on one scale, her utility is $\log(k)$, for example. Then, if her current wealth is £200, she shouldn't accept Bets 1' and 2' from above. After all, the expected utility of accepting Bet 1' is $0.3 \times \log(200 + 35 - 100) + 0.7 \times \log(200 + 35) \approx 5.293$, and that is less than the expected utility of refusing it, which is $\log(200) \approx 5.298$. And similarly for Bet 2'. However, it turns out that there are prices at which Finella should buy and sell £100 bets that will lead to a Dutch Book. Her credences require her to sell a £100 bet on *Low* for £36, since the expected utility of accepting that bet at that price exceeds the expected utility of refusing it. Likewise, they require her to sell a £100 bet on *Medium* for £47 and buy a £100 bet on *Low* ∨ *Medium* for £85. And, taken together, these bets are guaranteed to lose her £2.

However, consider Finella's colleague, Finch. His current wealth is also £200, and his utility function is also logarithmic in money. His credences are as follows:

| *Low* | *Medium* | *Low* ∨ *Medium* |
|---|---|---|
| 0.3 | 0.4 | 0.71 |

In some sense, Finch's violation of Finite Additivity is less extreme than Finella's because his credence in the disjunction exceeds the sum of his credences in the disjuncts by far less than in Finella's case. Now, there are no prices on £100 bets that Finch's logarithmic utility function permits him to accept that will also give rise to a Dutch Book. Even if he sells £100 bets on *Low* and *Medium* for the minimum prices that expected utility considerations will permit (£35.96 and £46.32, respectively), and even if he buys a £100 bet on *Low* ∨ *Medium* for the maximum they'll permit (£65.10), he still won't end up with a guaranteed loss.

This is a problem for the Dutch Book argument for Finite Additivity. Finch violates that law of credence, but we cannot find a book of £100 bets, each of which he is rationally required or even permitted to accept, and which, taken together, guarantee him a net loss. There are two putative solutions to this problem.

On the first, we specify bets not in terms of money but in terms of some other commodity such that (i) quantities of this commodity can be measured by real numbers, and (ii) our utility is a linear function of that commodity measured in that way – that is, the utility that you gain from obtaining a particular quantity of it is the same regardless of how much of it you have already.[5]

On the second putative solution, we note that, while Finch's utility is not linear in money, if it is a reasonably well-behaved, strictly increasing function of money, it will approximate being linear in money as closely as you could wish, so long as you restrict attention to a suitably narrow range of possible losses and gains (de Finetti, 1974, Section 3.2.6). We noted above that we cannot set prices on £100 bets that Finch's credences will require him to accept, but which will amount to a Dutch Book when taken together. The problem is that the segment of his utility function that starts with his utility in £100 and runs through to his utility in £300 does not sufficiently approximate linearity. However, suppose we consider £1 bets rather than £100 bets – that is, a bet that pays £1 if the proposition in question is true and £0 if it is false. So we are looking for a price at which Finch should sell £1 bets on *Low* and on *Medium*, and a price at which he should buy a £1 bet on their disjunction such that, taken together, these bets at these prices form a Dutch Book. And it turns out that such bets exist: Finch should sell a £1 bet on *Low* for £0.301; he should sell a £1 bet on *Medium* for £0.401; and he should buy a £1 bet on *Low* ∨ *Medium* for £0.703. This will end up losing him £0.001 for sure. And indeed this holds quite generally. That is, for any violation of Finite Additivity or any violation of Normalisation, for any

---

[5]  See Schick (1986) for scepticism about such a move.

level of wealth, and for any reasonably well-behaved utility function, there will be some bets and prices that you should accept that form a Dutch Book against you. And similarly for Regularity, Countable Additivity, and the Principal Principle.

As well as suggesting this solution, Ramsey also pointed to what he saw as a flaw in it (Ramsey, 1926 [1931], 176). Notice that, because her credence in *Low* ∨ *Medium* was so much higher than the sum of her credences in *Low* and *Medium*, we could make Finella suffer a loss of £2 for sure; but when it came to Finch's credences, where the difference between the credence in the disjunction and the sum of the credences in the disjuncts is so much less, we could only make him suffer a loss of £0.001 for sure. Ramsey worries that, since the latter amount is a mere trifle to Finch, being made to suffer that loss for sure might be a weak indication of irrationality. But it seems to me that this is a virtue of the account, not a vice. After all, Finch does seem to be less irrational than Finella. His violation of Finite Additivity seems less egregious. And it seems that the much smaller sure loss he can be made to suffer reflects this fact.

In what follows, I'll assume the first solution. In fact, I'll write as if our utility is linear in money. But money should be taken as a placeholder for such a commodity. If, however, there is no such commodity, I suggest we fall back on the second solution and note that, for small enough stakes, our utility approximates being linear in money as closely as you like.

## 3.4 Look before You Bet

Recall again Finella's credences:

| Low | Medium | Low ∨ Medium |
| --- | --- | --- |
| 0.3 | 0.4 | 0.9 |

The Dutch Book argument against her assumes that she should sell a £100 bet on *Low* for £35, sell a £100 bet on *Medium* for £45, and buy a £100 bet on *Low* ∨ *Medium* for £85. But, you might respond, while it is true that Finella should buy a £100 bet on *Low* ∨ *Medium* for £85 *if she has not engaged in any other bets*, surely she should *not* buy it for that price if she has already sold £100 bets on *Low* and *Medium* for £35 and £45, respectively. After all, Finella herself is just as able as we are to see that together these bets are guaranteed to lose her money. That is, if she is offered the bets in the order they are presented above, she will be able to see the sure loss coming when she is offered the final bet on the disjunction. So, in that situation, surely she should not accept the final bet. And indeed the same is true if the bets are offered in any other order. If Finella has accepted the first two, she will be able to see that accepting the third will lead her to a sure loss. So, contrary to the first premise of the Dutch Book argument, surely she is not rationally required to accept that final bet – indeed, surely she is rationally forbidden from doing so. In other words, it just isn't true that you should always pay £85 for a £100 bet on a proposition that you believe with credence 0.9. And the same applies for other credences and corresponding £100 bets.

All of this is a bit quick. Let's suppose Finella has sold the £100 bet on *Low* for £35 and the £100 bet on *Medium* for £45. She's then offered the £100 bet on *Low* ∨ *Medium* for £85. Her choices are these: accept the bet at that price, or refuse it. Let's look at the net gain of these two options at the three possible worlds:

|  | *Refuse* | *Accept* |
|---|---|---|
| *Low* | $£(35 - 100 + 45) = -£20$ | $£(35 - 100 + 45 - 85 + 100) = -£5$ |
| *Medium* | $£(35 + 45 - 100) = -£20$ | $£(35 + 45 - 100 - 85 + 100) = -£5$ |
| *High* | $£(35 + 45) = £80$ | $£(35 + 45 - 85) = -£5$ |

If she accepts, she is sure to end up with less than she had at the beginning of the whole process. If she refuses, she is not. But that does not entail that she should refuse.

Consider an analogous case. Suppose Brodie steals £20 from me. I find out and confront him, but neither of us is sure that my evidence is sufficient to convict him. Brodie offers to return £15 of my money to me if I don't report the theft to the police. What should I do? If I accept Brodie's offer, I'm sure to be £5 poorer than I was at the start of the process. If I don't and the police find Brodie guilty, I receive the £20 that Brodie stole as well as compensation of £80; if I don't accept Brodie's offer and the police don't find him guilty, I don't receive any of my money back, so I end up £20 poorer than I was at the start of the process. So, accepting Brodie's offer guarantees that I will lose money overall (£5); whereas if I go to the police, there's some chance I'll end up making money (£80), and some chance I'll end up losing money (£20). But it is not obvious that I should refuse Brodie's offer. When I make my decision, he has already stolen the money. That fact is fixed and my decision can't change it. I should choose in order to maximise my wealth against that fixed backdrop.

The same holds for Finella's bets. She has accepted the first two bets. That fact is fixed and her decision can't change it. She should choose now in order to maximise her wealth against that fixed backdrop. She is offered a £100 bet on *Low* ∨ *Medium* for the price of £85. She should appeal to her credence in that disjunction to make the choice. Indeed, her choice is really between a new bet and a different sure thing. She is choosing between a loss of £5 for sure (if she accepts the final bet of the three) and a bet that loses her £20 if *Low* or *Medium* is true, and gains her £80 if *High* is true (if she refuses it). If she calculates the expectation of this new bet, it is £((0.9 × −20) + (0.1 × 80)) = −£10. Thus, it has a lower expected payout than the sure loss of £5. So she should take that sure loss – that is, she should accept the bet, just as (DB1), the first premise of the Dutch Book argument, says. So there's a sense in which whether or not Finella should accept the final bet in the Dutch Book depends not only on her credences and utilities, but also on the previous bets she's made – those previous bets fix the backdrop against which she makes her final decision. But even taking that into account, she should accept the final bet.

## 3.5 Possible Worlds

As we have formulated them so far, the Dutch Book arguments for Normalisation and Finite and Countable Additivity claim that, if you violate either of those laws, your credences require that you enter into each of a series of bets that, when taken together, result in a *sure loss*; that is, you are *guaranteed* to lose money by entering into these bets; that is, *however the world turns out*, you'd have been better off refusing all of the bets. But we haven't yet made clear what we mean by the italicised terms; we haven't specified the modality to which they appeal. And, as the examples below show, it is important to do so.

First example: Southey is 10 per cent confident in the proposition *I*, which says that Charlotte and Currer are the same person. Now, in fact, *I* is true. And so it is metaphysically necessary. But it is not logically necessary. And it is not epistemically necessary either, providing Southey does not have definitive evidence that Charlotte and Currer are the same person. His credence demands that he sell a £100 bet on *I* for £45 (Bet 1). The payout is as follows:

|  | Bet 1 | Net gain |
|---|---|---|
| $I$ | $-100 + 45$ | $-55$ |
| $\bar{I}$ | $45$ | $45$ |

The second line represents a metaphysical impossibility. So, if the modality relevant to Dutch Book arguments is metaphysical modality, Southey can be dutchbooked for this violation of the relevant version of Normalisation, but if it is logical or epistemic modality, he cannot be.

Here's a different sort of case: Hina is 80 per cent confident that it will rain tomorrow, but only 90 per cent confident that she is 80 per cent confident that it will rain tomorrow – that is, she isn't certain that her credences are as they in fact are. Let $R$ be the proposition that it will rain tomorrow, and let $C$ be the proposition that Hina is 80 per cent confident that it will rain tomorrow. Then her credences demand that she sell a £100 bet on $C$ for £95 (Bet 1) and buy a £1 bet on $R$ for 75p (Bet 2). Here are the payouts:

|  | *Bet 1* | *Bet 2* | Total |
|---|---|---|---|
| $R \& C$ | $1 - 0.75$ | $-100 + 95$ | $-4.75$ |
| $R \& \overline{C}$ | $1 - 0.75$ | $95$ | $95.25$ |
| $\overline{R} \& C$ | $-0.75$ | $-100 + 95$ | $-5.75$ |
| $\overline{R} \& \overline{C}$ | $-0.75$ | $95$ | $94.25$ |

Thus, if we include among the possible worlds at which the net gain of the bets is evaluated only those in which Hina's credences are as they in fact are – that is, only those worlds in which $C$ is true – these bets will lose Hina money at all possible worlds. That is, Hina will be deemed irrational just because she is not certain her credences are as they actually are. What's more, there is a seemingly plausible argument in favour of restricting the possible worlds in this way. After all, in worlds in which Hina's credences are different, her credences will require her to accept different bets. So Hina cannot mount a defence against the charge of irrationality by saying that, if her credence in rain had been different, and $C$ had been false, she would have made a net gain of £95.25 or £94.25. If her credences had been different, and $C$ had been false, she might not have been required to accept the bets in the first place! That is, in all worlds in which she has the credences that require her to take the bets, she makes a net loss – and that renders her irrational. Or so goes the argument. We'll address it fully below.

Let $\mathcal{W}$ be the set of possible worlds that is relevant for Dutch Book arguments. That is, $\mathcal{W}$ is the smallest set of possible worlds such that, if your credences demand that you enter into each of a series of bets that, taken together, lead to a net loss in all worlds in $\mathcal{W}$, then you are irrational. Which are the worlds in $\mathcal{W}$? All and only the metaphysically possible worlds? All and only the logically possible or epistemically worlds? All and only the metaphysically possible worlds in which your credences are as they in fact are at the actual world?

I propose the following: $W$ is the set of all and only the worlds that are epistemically possible for you at the time in question. Thus, Southey is not dutchbookable in either of the two examples above, for while it is not metaphysically possible for Charlotte and Currer to be different people, it is epistemically possible for Southey, and thus a world in which $I$ is false is an epistemically possible world for him. On my proposal, Hina is not dutchbookable either. While it is actually true that she has a credence of 80 per cent in rain – and so $C$ is actually true – it is at least epistemically possible from her point of view that she doesn't – and so there are epistemically possible worlds at which $C$ is false. In such worlds, Hina makes a net gain.

Why is $W$ the set of epistemically possible worlds? The simple answer is that we are using evaluations of the net losses or gains of bets at the various worlds in $W$ in order to assess a person's *rationality*. If I point out that Southey's credences demand that he enter into bets that will lose him money at all metaphysically possible worlds, you might reasonably defend him by pointing out that he doesn't *know* that those are the only metaphysically possible worlds. There are other worlds – worlds that are, in fact, metaphysically impossible – that, for all his evidence tells him, might be metaphysically possible, such as worlds at which Charlotte and Currer are different people. The fact that he will lose money at all metaphysically possible worlds does not, on its own, impugn his rationality. That is, in order to establish that someone is irrational, we have to show that their credences demand that they accept a series of bets that loses them money at every world that *for all they know* is the actual world.

As we saw, on my proposal, Hina's credences are not dutchbookable and thus not irrational. But what of the argument above, which noted that, in the epistemically possible worlds in which $C$ is false and Hina makes a net gain, her credences might not have required her to accept the pair of bets that is sure to lose her money if $C$ is true? Why does this not show that Hina is irrational for having credence 0.9 in $C$? After all, surely a mental state is irrational if having it requires you to make decisions that are sure to be detrimental to you *given that you are in that mental state*.

Actually, I think not. To see why, consider again the case of Southey's credence of 10 per cent in $I$, the proposition that Charlotte and Currer are the same person. Southey's credence requires him to make a decision – selling the £100 bet on $I$ for £45 – that is sure to be detrimental to him. After all, it loses him money in any metaphysically possible world. The fact he doesn't know this doesn't stop it from being the case. Therefore, *a fortiori*, that bet is sure to be detrimental to him *given that he has the mental states that he does have*. If the general principle above were true, therefore, Southey would be irrational, for he is in a mental state that requires him to make decisions that are sure to be detrimental to him, and therefore sure to be detrimental to him given that he is in that mental state. But we don't think that this renders him irrational. Doomed, perhaps, but not irrational, for he does not have access to the fact that his mental state dooms him. And the same holds in Hina's case.

So, we conclude that $\mathcal{W}$, the set of worlds relevant to Dutch Book arguments, is the set of all worlds that are epistemically possible for the individual whose credences are being assessed. We represent your epistemically possible worlds by an assignment of truth values to the propositions to which you assign credences. In particular, the worlds that are epistemically possible for you are precisely those represented by truth value assignments with the following features: (i) they are logically consistent, so that, if $\top$ is logically necessary, $\top$ is true, and if $A$ or $B$ or both are true, then $A \vee B$ is true, and otherwise it is false; (ii) the proposition that encodes your total evidence is true. Suppose $\mathcal{F}$ is the set of propositions to which you assign credences, and suppose $E$ is a proposition in $\mathcal{F}$ and $E$ encodes your total evidence. Then let $\mathcal{W}_{\mathcal{F}}^{E}$ be the set of truth value assignments to the propositions in $\mathcal{F}$ that represent the worlds that are epistemically possible for you.[6]

## 3.6 The Arguments Spelled Out

Having identified the set of worlds relevant to Dutch Book arguments in the previous section, we now have precise versions of our five laws from above. Suppose $\mathcal{F}$ is the set of propositions to which you assign credences. And suppose $E$ is in $\mathcal{F}$ and $E$ encodes your total evidence. Then:

> **Normalisation$_{\mathcal{F}}^{E}$** If $\top$ is in $\mathcal{F}$, and $\top$ is true at all worlds in $\mathcal{W}_{\mathcal{F}}^{E}$, then your credence in $\top$ should be maximal (i.e. 1).

> **Finite Additivity$_{\mathcal{F}}^{E}$** If $A$, $B$, and $A \vee B$ are all in $\mathcal{F}$, and there is no world in $\mathcal{W}_{\mathcal{F}}^{E}$ at which $A$ and $B$ are both true, then your credence in $A \vee B$ should be the sum of your credence in $A$ and your credence in $B$.

---

[6] Note that, on this account, no logical falsehood is epistemically possible. This might seem too strong. After all, before I learned of Wiles' proof in 1995, surely both Fermat's Last Theorem and its negation were epistemically possible for me, yet the latter was a logical falsehood. It's an interesting question how Dutch Book arguments might work without this assumption, but I won't address it here.

**Countable Additivity**$_{\mathcal{F}}^{E}$ If $A_1, A_2, \ldots$, and their infinite disjunction $A_1 \vee A_2 \vee \ldots$ are all in $\mathcal{F}$, and, for any $i \neq j$, there is no world in $\mathcal{W}_{\mathcal{F}}^{E}$ at which $A_i$ and $A_j$ are both true, then your credence in $A_1 \vee A_2 \vee \ldots$ should be the sum of your credences in $A_1, A_2, \ldots$.

**Regularity**$_{\mathcal{F}}^{E}$ If there is a world in $\mathcal{W}_{\mathcal{F}}^{E}$ at which $A$ is true, then your credence in $A$ should be positive.

**Principal Principle**$_{\mathcal{F}}^{E}$ If $A$ & $C_r^A$ and $C_r^A$ are both in $\mathcal{F}$ and your credence in $C_r^A$ is positive, then your credence in $A$ conditional on $C_r^A$ should be $r$.

With these in hand, we can spell out the Dutch Book arguments for these laws. Each shares the following first premise, which we introduced in Section 3.2:

(DB1)  If you have credence $p$ in proposition $A$, then

(i)   you are rationally permitted to pay £$x$ for a £$S$ bet on $A$ for any $x \leq pS$,
(ii)  you are rationally required to pay £$x$ for a £$S$ bet on $A$ for any $x < pS$.

In each case, the second premise is a mathematical theorem. To state these theorems, we need some terminology. There are three features of a Dutch Book that we must specify:

- first, the number of bets it comprises – a single bet (*singly exploitable*), finitely many bets (*finitely exploitable*), or countably many bets (*countably exploitable*)?
- second, the normative status of the bets relative to the individual's credences – rationally required (*strongly exploitable*) or rationally permitted (*weakly exploitable*)?
- third, the undesirable feature of the payoff – net loss in every world (*fully exploitable*), net loss in some worlds and net gain in none (*almost exploitable*), or expected net loss relative to all possible chance functions (*exploitable in expectation*)?

So, for instance, you are *singly, strongly, fully* exploitable iff there is a *single* bet that your credences *require* you to accept that *loses you money in all* epistemically possible worlds. These are generalisations of de Finetti's notion of incoherence, which we mentioned above. You are incoherent, in de Finetti's sense, if you are finitely, strongly, fully exploitable.

Then we have the Dutch Book Theorems for Normalisation (N), Finite (FA) and Countable Additivity (CA), Regularity (R), and the Principal Principle (PP):

**Theorem 3.1**   *Suppose (DB1) holds. Then:*

(DB2N)
(DB2FA)
(DB2CA)
(DB2R)
(BD2PP)

*If your credences violate* $\left\{\begin{array}{l} N \\ FA \\ CA \\ R \\ PP \end{array}\right.$ *then they are* $\left\{\begin{array}{l} \textit{singly strongly fully exploitable} \\ \textit{finitely strongly fully exploitable} \\ \textit{countably strongly fully exploitable} \\ \textit{singly weakly almost exploitable} \\ \textit{finitely strongly exploitable in expectation} \end{array}\right.$

These make up the second premises of their respective arguments. And finally we have the third premises:

(DB3N)
(DB3FA)
(DB3CA)
(DB3R)
(DB3PP)

If your credences are $\left\{\begin{array}{l} \text{singly strongly fully exploitable} \\ \text{finitely strongly fully exploitable} \\ \text{countably strongly fully exploitable} \\ \text{singly weakly almost exploitable} \\ \text{finitely strongly exploitable in expectation} \end{array}\right.$ then they are irrational.

So the five arguments run as follows:

| Normalisation | Finite Additivity | Countable Additivity | Regularity | Principal Principle |
|---|---|---|---|---|
| (DB1) | (DB1) | (DB1) | (DB1) | (DB1) |
| (DB2N) | (DB2FA) | (DB2CA) | (DB2R) | (DB2PP) |
| (DB3N) | (DB3FA) | (DB3CA) | (DB3R) | (DB3PP) |

## 3.7 Countable Exploitability

In this section, we consider two styles of objections to (DB3CA), the third premise of the Dutch Book argument for Countable Additivity, which says that countably exploitable credences are irrational. One is originally due to de Finetti (1972, 91) and expanded by Vann McGee (1999), the other to Frank Arntzenius et al. (2004). I judge these to be decisive objections, and this will lead us to abandon this argument.

Let's start with the sort of objection that de Finetti and McGee raise. Each describes a set of credences that are intuitively rational, but which are countably exploitable. In de Finetti's argument, we imagine that a natural number has been chosen at random. For $i = 1, 2, 3, \ldots$, let $N_i$ be the proposition that $i$ was the number chosen. De Finetti claims that it must be rationally permissible to assign the same credence to each $N_i$, and credence 1 to their infinite disjunction, $N$, which says that some natural number was chosen. But the only probabilistic credences that do this assign 0 to each $N_i$, and 1 to $N$, and that violates Countable Additivity, since an infinite sum of 0s is 0. So these intuitively rational credences are countably exploitable, and countable exploitability cannot entail irrationality.

Next, we present McGee's version. He describes a set of intuitively rational credences that satisfy Countable Additivity but are nonetheless countably exploitable. Let $M$ be the proposition that Earth's global mean surface temperature will rise by more than 0°C and less than 1°C. Let $M_1$ be the proposition that the increase will be more than 0°C and less than 0.5°C; let $M_2$ be the proposition that the increase will be at least 0.5°C and less than 0.75°C; and so on. And consider Kanat, whose credences are these: in $M$ is $\frac{1}{2}$, in $M_1$ is $\frac{1}{4}$, in $M_2$ is $\frac{1}{8}$, in $M_3$ is $\frac{1}{16}$, and so on. So the sum of his credences in $M_1, M_2, \ldots$ is $\frac{1}{4} + \frac{1}{8} + \frac{1}{16} + \ldots = \frac{1}{2}$, which is his credence in their infinite disjunction, $M$ (see Figure 3.1).

Now, Kanat's credences satisfy Countable Additivity. And we might assume they satisfy Normalisation and Finite Additivity as well. What's more,

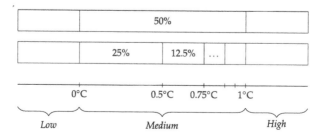

**Figure 3.1** Kanat's credences.

intuitively, we judge Kanat's credences rationally permissible. For instance, suppose Kanat's evidence strongly supports a credence of $\frac{1}{2}$ that temperatures will rise by between 0°C and 1°C, but provides nothing that allows him to tell between the different more specific temperature rises within that range. In that case, intuitively, his credences provide a rational response, and indeed possibly the unique rational response, to this evidence.

Now, suppose Kanat is faced with an infinite series of decision problems. A *decision problem* consists of a set of available acts between which you must choose. For instance, when you are offered a bet for a certain price, you face a decision problem in which the available acts are accepting the bet at that price and rejecting it. A bet on a proposition is an act that has two possible outcomes – one when the proposition is true and one when it is false. But sometimes there are available acts with more than two outcomes, and the decision problems that Kanat faces include acts like this. In the first decision problem in our infinite series, Kanat must choose between acts $a_1$ and $b_1$; in the second, between $a_2$ and $b_2$; and so on. Each act $b_i$ has a pay-off of £0 for sure. The pay-offs of each act $a_i$ are given as follows (in pounds):

$$a_1 = \begin{cases} 7 & \text{if } M_1 \\ -1 & \text{otherwise} \end{cases} \qquad a_i = \begin{cases} -i2^i & \text{if } M_{i-1} \\ (i+1)2^{i+1} & \text{if } M_i \\ 0 & \text{otherwise} \end{cases}$$

Now, (DB1), the first premise of each of the Dutch Book arguments we've considered so far, tells us how to choose when faced with a decision between accepting a bet and rejecting it. But, for $i \geq 2$, the choice between $a_i$ and $b_i$ is not such a decision, since $a_i$ has three outcomes. So how should we choose? Well, we have assumed that Kanat satisfies Normalisation and Finite Additivity – after all, we have the less contentious Dutch Book arguments for those laws, and (DB1) is sufficient to get them going. So, we might appeal to orthodox decision theory when we make our choice. That is, we might say that, faced with the decision problems $a_1$ versus $b_1$, $a_2$ versus $b_2$, and so on, Kanat should choose by maximising expected utility.

Let's see what this means. A decision problem is specified by the set of available acts between which any agent who faces it must choose. An act is specified by its outcome at each possible world. Given each act, we calculate its expected utility relative to our credence function and our utility function. That is, we take each possible outcome of the act, look at the utility we assign to that outcome, and we weight that utility by the credence we assign to that outcome if we choose that act. And then we add up these credence-weighted utilities to give our expected utility for the act. Thus, if $c$ is our credence function and $u$ is our utility function, and $a$ is an act and $\mathcal{O}$ is the set of possible outcomes of our acts, then let $a \Rightarrow o$ be the proposition that says that $o$ results from choosing $a$. Thus, $a \Rightarrow o$ is true at exactly those worlds at which the outcome of $a$ is $o$. Then $u(o)$ is our utility for $o$ and $c(a \Rightarrow o)$ is our credence that $o$ will result from choosing $a$. So, the expected utility of $a$ relative to $c$ and $u$ is:

$$EU_{c,u}(a) = \sum_{o \in O} c(a \Rightarrow o)u(o)$$

Thus, for example, consider act $a_2$, which pays out $-£8$ if $M_1$ is true, $£24$ if $M_2$ is true, and $£0$ otherwise. Then

$$EU_{c,u}(a_2) = c(M_1)u(-£8) + c(M_2)u(£24) + c(\overline{M_1 \vee M_2})u(£0)$$
$$= \frac{1}{4}(-8) + \frac{1}{8}(24) + \frac{5}{8}(0) = 1$$

The law of decision called *Maximise Expected Utility*, which we will abbreviate *MEU*, then tells you to choose an act with maximal expected utility relative to your credence and utility functions.

Now, notice first that, in fact, each $a_i$ has an expected utility of 1, while each $b_i$ has an expected utility of 0.[7] So, if Kanat maximises expected utility, he should pick $a_1$ over $b_1$, $a_2$ over $b_2$, and so on. But here's the rub: whichever of $M_i$ is true, choosing the infinite series of acts $a_1$ & $a_2$ & ... results in less money than choosing the series $b_1$ & $b_2$ & .... Indeed, the former series of choices is sure to result in a loss of £1, as the following table illustrates:

|  | $a_1$ | $a_2$ | $a_3$ | $a_4$ | $a_5$ | ... | Total |
|---|---|---|---|---|---|---|---|
| $\overline{M}$ | $-1$ | 0 | 0 | 0 | 0 | ... | $-1$ |
| $M_1$ | 7 | $-8$ | 0 | 0 | 0 | ... | $-1$ |
| $M_2$ | $-1$ | 24 | $-24$ | 0 | 0 | ... | $-1$ |
| $M_3$ | $-1$ | 0 | 64 | $-64$ | 0 | ... | $-1$ |
| $M_4$ | $-1$ | 0 | 0 | 160 | $-160$ | ... | $-1$ |
| ⋮ | ⋮ | ⋮ | ⋮ | ⋮ | ⋮ | ⋮ | ⋮ |

---

[7] Here's the calculation:

• The expected value of $a_1$ is

$$EU(a_1) = \left(\frac{1}{4} \times 7\right) + \left(\frac{3}{4} \times -1\right) = 1;$$

• For $i > 1$, the expected value of $a_i$ is

$$EU(a_i) = \left(\frac{1}{2^i} \times (-i2^i)\right) + \left(\frac{1}{2^{i+1}} \times (i+1)2^{i+1}\right) + \left(\left(1 - \frac{1}{2^i} - \frac{1}{2^{i+1}}\right) \times 0\right) = 1.$$

The latter series, on the other hand, pays out £0 for sure. Thus, Kanat is countably exploitable. But, as we noted above, Kanat's credences are eminently reasonable. So this example, adapted from McGee (1999), seems to show that being countably exploitable is no marker of irrationality.

This conclusion is further supported by a slightly different objection to (DB3CA), due to Arntzenius et al. (2004). (DB3CA) claims that, if, faced with a countable series of decision problems, your credences lead you to make each of a series of choices that is dominated by an alternative series of choices, then those credences are irrational. Our conviction that this is true is based on a more fundamental belief that it is irrational to choose each of a dominated series of actions, whether that series is finite or infinite – your credences are irrational because they require you to behave in a way that is irrational. But the following example, adapted from Arntzenius et al. (2004), shows this isn't so.

Satan chops an apple into countably many pieces and offers them to Eve. If she's only eaten finitely many pieces of apple so far, eating any further piece of apple adds some positive utility to Eve's existing utility; but if she eats them all, she is expelled from Eden, and she assigns negative utility to that. She is offered each piece of apple in turn. Clearly, she is rationally required to take it rather than refuse it, since taking it increases her utility for sure, while refusing it leaves her utility as it is. But she is rationally required not to take them all, since that has negative utility, while refusing them all has zero utility. That is, it is simply not the case that you are required to perform the infinite conjunction of a set of actions each of which you are required to perform. And this undermines (DB3CA), and thus the Dutch Book Argument for Countable Additivity. In what follows, then, we won't pursue this argument.

## 3.8 The Converse Dutch Book Theorem

Consider Fiona. She assigns credences only to two propositions, *Low* and *Low* ∨ *Medium*. Her credence in the first is 0.5 and in the second 0.3. Then her credences require her to buy a £100 bet on *Low* for £45 and sell a £100 bet on *Low* ∨ *Medium* for £35. Taken together, these bets lose her money for sure. That is, her credences are finitely strongly fully exploitable. However, she satisfies Finite Additivity. To violate Finite Additivity, you must assign credences to at least three propositions, $A$, $B$, and $A ∨ B$, where you know $A$ and $B$ are not both true. And your credence in $A ∨ B$ must differ from the sum of your credence in $A$ and your credence in $B$. But Fiona assigns credences only to two propositions, *Low* and *Low* ∨ *Medium*. And the same applies for Normalisation. Suppose Norman assigns credences only to *Low* and *Medium* ∨

*High*, both of which are epistemically possible. He assigns 0.2 to both. Then he'll sell a £100 bet on each for £25. Taken together, these bets lose him money for sure. So he is finitely strongly fully exploitable, but since he doesn't assign a credence to any epistemically necessary proposition, he doesn't violate Normalisation.

So Probabilism, the conjunction of Normalisation and Finite Additivity, is not the strongest law of credence we can establish by appealing to finite strong full exploitability – there are credences that violate neither that are finitely strongly fully exploitable. So, what is the strongest such law? Similar questions arise for finite weak almost exploitability and finite strong exploitability in expectation. We can identify credences that satisfy Regularity because they assign only positive credences, but are nonetheless finitely weakly almost exploitable. For instance, this will be the case if (i) you assign credences only to $A$ and to $A \vee B$, (ii) you assign 0.6 to both, and (iii) $\overline{A}$ & $B$ is epistemically possible. And we can identify credences that satisfy the Principal Principle because they don't assign a credence to any chance hypotheses, but are nonetheless finitely strongly exploitable in expectation. For instance, this will be the case if (i) you assign a credence only to $A$, (ii) you assign 0.4 to that, and (iii) the only epistemically possible objective chance for $A$ is 0.6.

## 3.9 The Full Versions of the Laws

Fiona avoids violating Finite Additivity just by failing to assign a credence to *Medium*; if she were to assign a credence to that proposition, she'd be guaranteed to fall foul of Finite Additivity. And Norman avoids violating Normalisation by failing to assign a credence to *Low* ∨ *Medium* ∨ *High*; if he were to assign a credence to that proposition, he'd violate either Normalisation or Finite Additivity. Our strengthened versions of the laws demand that it must be possible to extend your credences by assigning new credences to certain missing propositions in such a way that these extended credences satisfy the original law. We call these the *full versions* of the laws. While Fiona satisfies the original version of Finite Additivity, she violates its full version. But which are the missing propositions? Well, that depends on the law in question.

Let's take Normalisation, Finite Additivity, and Regularity first. Indeed, instead of considering the norms individually, we'll consider certain combinations: Normalisation + Finite Additivity, which we call Probabilism; and Probabilism + Regularity. Suppose $\mathcal{F}$ is a set of propositions. Then we say that *$\mathcal{F}$ is an algebra* if it is closed under the Boolean operators of negation (¬), disjunction ($\vee$), and conjunction (&).[8] Given a set of propositions $\mathcal{F}$, let $\mathcal{F}^*$ be the smallest algebra that contains $\mathcal{F}$.[9] Then the new version of Probabilism says:

> **Full Probabilism (FP)** Suppose $\mathcal{F}$ is the set of propositions to which you assign credences, and $E$ is the strongest proposition you have as evidence. Then: it should be possible to extend your credences over $\mathcal{F}$ to credences over $(\mathcal{F} \cup \{E\})^*$, so that the extended credences satisfy
>
> (i)   Normalisation$^E_{(\mathcal{F} \cup \{E\})^*}$ and
> (ii)  Finite Additivity$^E_{(\mathcal{F} \cup \{E\})^*}$.

And the new version of Probabilism + Regularity says:

> **Full Probabilism + Regularity (FPR)** Suppose $\mathcal{F}$ is the set of propositions to which you assign credences, and $E$ is the strongest proposition you have as evidence. Then: it should be possible to extend your credences over $\mathcal{F}$ to credences over $(\mathcal{F} \cup \{E\})^*$, so that the extended credences satisfy
>
> (i)   Normalisation$^E_{(\mathcal{F} \cup \{E\})^*}$,
> (ii)  Finite Additivity$^E_{(\mathcal{F} \cup \{E\})^*}$, and
> (iii) Regularity$^E_{(\mathcal{F} \cup \{E\})^*}$.

The combination of Probabilism + Principal Principle requires a little more. Not only must we add $E$ to $\mathcal{F}$ and then close under negation, disjunction, and conjunction. We must also add propositions that specify the epistemically possible hypotheses about the objective chances. Suppose $ch$ is a probability function. Then let $C_{ch}$ be the proposition that says that $ch$ gives the objective chances. We make the following crucial and substantial assumption: for each possible objective chance function $ch$, $ch(C_{ch}) = 1$ (see footnote 4). With this in hand, we can offer a slightly strengthened version of the Principal Principle. Instead of demanding only that $c(A|C_r^A) = r$ whenever $c(C_r^A) > 0$, we demand further that $c(A|C_{ch}) = ch(A)$ whenever $c(C_{ch}) > 0$. The latter entails the former, but not vice versa. We'll call this the Strong Principal Principle. Now, to begin

---

[8]  That is,

  (i)  whenever $A$ is in $\mathcal{F}$, $\neg A$ is in $\mathcal{F}$;
  (ii) whenever $A$ and $B$ are both in $\mathcal{F}$, $A \vee B$ and $A$ & $B$ are both in $\mathcal{F}$.

[9]  That is,

  (i)   $\mathcal{F}^*$ is an algebra;
  (ii)  $\mathcal{F} \subseteq \mathcal{F}^*$;
  (iii) if $\mathcal{F}^\dagger$ is an algebra and $\mathcal{F} \subseteq \mathcal{F}^\dagger$, then $\mathcal{F}^* \subseteq \mathcal{F}^\dagger$.

with, we assume that, as for Pritpal, there are only finitely many epistemically possible chance functions.

> **Full Probabilism + Strong Principal Principle** Suppose $\mathcal{F}$ is the set of propositions to which you assign credences, $E$ is the strongest proposition you have as evidence, and $\mathcal{C} = \{ch_1, \ldots, ch_n\}$ is the set of epistemically possible chance functions. Then: it should be possible to extend your credences over $\mathcal{F}$ to credences over $(\mathcal{F} \cup \{C_{ch_1}, \ldots, C_{ch_n}, E\})^*$, so that the extended credences satisfy
>
> (i)   Probabilism$^E_{(\mathcal{F} \cup \{C_{ch_1}, \ldots, C_{ch_n}, E\})^*}$ and
> (ii)  the Strong Principal Principle$^E_{(\mathcal{F} \cup \{C_{ch_1}, \ldots, C_{ch_n}, E\})^*}$, which says that, if $c(C_{ch}) > 0$, then $c(A|C_{ch}) = ch(A)$.

But what if there are infinitely many epistemically possible chance functions? In that case, our law of credence looks a little different. And it requires some new pieces of terminology. Suppose $\mathcal{Z}$ is a set of credence functions. Then:

- We say that $\mathcal{Z}$ *is convex* if, whenever it contains two credence functions $c$, $c'$, it also contains all their mixtures $\lambda c + (1 - \lambda)c'$.
- We say that $\mathcal{Z}$ *is closed* if, for any infinite sequence $c_1, c_2, \ldots$ of credence functions in $\mathcal{Z}$ that converges to a limit $c$, $c$ is also in $\mathcal{Z}$.
- The convex hull of $\mathcal{Z}$ is the smallest convex set that contains all the credence functions in $\mathcal{Z}$. We denote it $\mathcal{Z}^+$.
- The closure of $\mathcal{Z}$ is the smallest closed set that contains all the credence functions in $\mathcal{Z}$. We denote it $cl(\mathcal{Z})$.

> **Full Probabilism + Strong Principal Principle$^+$** Suppose $\mathcal{F}$ is the set of propositions to which you assign credences, and $\mathcal{C}$ is the set of epistemically possible chance functions. Then: your credence function should lie within the closure of the convex hull of $\mathcal{C}$ – that is, it should lie in $cl(\mathcal{Z}^+)$.

## 3.10 The Dutch Book Argument for the Full Norms

We can now rewrite our earlier arguments for the laws we are considering. Following are the new Dutch Book arguments for Full Probabilism (FP), Full Probabilism + Regularity (FPR), Full Probabilism + Strong Principal Principle (FPPP), and Full Probabilism + Strong Principal Principle$^+$ (FPPP$^+$).

**Theorem 3.2** *Suppose (DB1) holds. Then:*

$$
\begin{array}{l}
\text{(DB2FP)} \\
\text{(DB2FPR)} \\
\text{(DB2FPPP)}
\end{array}
\quad \textit{Your credences violate} \left\{ \begin{array}{l} FP \\ FPR \\ FPPP \end{array} \right\} \textit{iff they are} \left\{ \begin{array}{l} \textit{finitely strongly fully exploitable} \\ \textit{finitely weakly almost exploitable} \\ \textit{finitely strongly exploitable in expectation} \end{array} \right.
$$

We sketch the proof in Section 8. Finally, we have the third premises:

$$
\begin{array}{l}
\text{(DB3FP)} \\
\text{(DB3FPR)} \\
\text{(DB3FPPP)}
\end{array}
\quad \text{If your credences are} \left\{ \begin{array}{l} \text{finitely strongly fully exploitable} \\ \text{finitely weakly almost exploitable} \\ \text{finitely strongly exploitable in expectation} \end{array} \right.
$$

and there are credences that are not $\left\{ \begin{array}{l} \text{finitely strongly fully exploitable} \\ \text{finitely weakly almost exploitable} \\ \text{finitely strongly exploitable in expectation} \end{array} \right.$

then your credences are irrational.

So the three arguments run as follows:

| (DB1) | (DB1) | (DB1) |
|---|---|---|
| (DB2FP) | (DB2FPR) | (DB2FPPP) |
| (DB3FP) | (DB3FPR) | (DB3FPPP) |
| FP | FPR | FPPP/FPPP$^{+}$ |

# 4 Updating and Evidence

So far in this Element, we have considered only synchronic laws for credences. Normalisation and Regularity, Finite and Countable Additivity, and the Principal Principle all specify how your credences at a given time should relate to one other. In this section, we ask about diachronic laws. These tell you how your credences at one time should relate to your credences at another. In particular, we'll be interested in laws that purport to govern how you should update your credences in response to new evidence.

Roughly speaking, there are two sorts of updating law: the first governs *actual* features of your updating behaviour, such as the relationship between your prior credences and the credences you *actually* adopt after receiving the evidence you *actually* receive; the second governs *both actual and counterfactual* features of your updating behaviour, such as the relationship between your prior credences and the credences you *would* or *might* adopt were you to learn one thing, and the credences you *would* or *might* adopt were you to learn some other thing. As we will see, there are no good sure loss arguments for the first sort of law.[10] We'll see this in two ways. First, while there's one sort of sure loss to which you are vulnerable if you actually don't update in the prescribed way, you are also vulnerable to this sort of loss if you actually do update in the prescribed way. Second, we'll see that, while there's a second sort of sure loss that doesn't have the problems of the first, it is possible actually to update in the prescribed way and yet still be vulnerable to this sort of sure loss (if you wouldn't have updated in the prescribed way had you learned something else), and it is possible to not actually update in the prescribed way and yet nonetheless not be vulnerable to this sort of sure loss (if you might have updated differently in the light of the evidence you actually received, and if those other possible updates differ from your actual update in a certain way). Thus, by the end of the first section, we'll conclude that sure loss arguments for updating laws target irreducibly modal features of your updating – they target the relationship between your prior credences and your *updating rule*, rather than between your prior credences and your *actual updating behaviour*. We'll then turn to an objection to such arguments and see why it fails.

A note on terminology: So far, we have been talking about sets of credences, or assignments of credences to sets of propositions, but we haven't used any notation to describe these. In this section, it will be helpful to have such notation. Suppose an individual assigns credences to all and only the propositions

---

[10] Brian Skyrms has consistently raised this point (e.g. Skyrms, 1993). What we will call *updating rules*, Skyrms calls *epistemic strategies*.

in the set $\mathcal{F}$. Then her *credence function* is the function $c : \mathcal{F} \to [0, 1]$ such that, for each $A$ in $\mathcal{F}$, $c(A)$ is her credence in $A$. We say that a credence function is probabilistic if it satisfies Full Probabilism.

## 4.1 Diachronic Dutch Books and Actual Conditionalisation

To illustrate our points, we'll pursue an example throughout the section. Consider Conrad. Like Pritpal from the beginning of Section 2, he is wondering about the landfall of Storm Saoirse. He knows it will land first in one of the following locations: Stranraer, St Davids, Silloth, or Southport. Let *England* (or $E$) be the proposition that it lands in England – that is, Silloth or Southport. And let *North* (or $N$) be the proposition that it lands north of the Lake District – that is, Stranraer or Silloth. Here are Conrad's credences on Monday:

| Stanraer | St Davids | Silloth | Southport | North | England |
|----------|-----------|---------|-----------|-------|---------|
| 10%      | 40%       | 30%     | 20%       | 40%   | 50%     |

On Tuesday, Conrad learns *England* (i.e. Silloth or Southport), and he learns nothing further. In the light of this, he updates his credences. He now has credence 70 per cent in *North* (i.e. Silloth or Stranraer). We judge him irrational. If he were rational, upon learning *England* he would remove all of his credence from St Davids and Stranraer and redistribute it among Silloth and Southport in proportion to his Monday credences in those possibilities. Thus, he would have credence 60 per cent in Silloth and 40 per cent in Southport, and thus credence 60 per cent in *North*. That is,

| Stanraer | St Davids | Silloth | Southport | North | England |
|----------|-----------|---------|-----------|-------|---------|
| 0%       | 0%        | 60%     | 40%       | 60%   | 100%    |

But he doesn't.

So Conrad violates a putative law that governs his actual updating behaviour:

**Actual Conditionalisation** If

(i)   $c$ is your credence function at $t$ (this is often known as your *prior credence function* or just your *prior*),

(ii)  $c'$ is your credence function at a later time $t'$ (this is often known as your *posterior credence function* or just your *posterior*),

(iii) $E$ is the strongest evidence you obtain between $t$ and $t'$, and

(iv)  $c(E) > 0$,

then it ought to be that

(v)    $c'(A) = c(A|E)$, for all $A$ in $\mathcal{F}$.[11]

Is there a sure loss argument against Conrad? As we said at the beginning, there is one sort of sure loss argument against him, but this proves too much. There is also another sort of sure loss argument, but it does not apply to Conrad only on the basis of his actual updating behaviour – if it applies to him at all, it is on the basis of other modal features of him.

By the first premise of our Dutch Book arguments, (DB1), Conrad's prior credences on Monday require him to sell a £100 bet on *North* for £45, since he's 40 per cent confident in *North* on Monday. And his posterior credences on Tuesday require him to buy a £100 bet on *North* for £55, since he's 70 per cent confident in *North* on Tuesday. Taken together, these two bets lose Conrad £10 in all epistemically possible worlds. Is this diachronic Dutch Book sufficient to establish that Conrad is irrational?

Let's hope not! After all, Conrad would also have been required to accept that same pair of bets on Monday and Tuesday had he conditionalised on the evidence he learned, rather than updating in the way he did. Had he conditionalised, his credences on Monday would have been the same, so they'd have required him to sell a £100 bet on *North* for £45. And his credence in *North* on Tuesday would have been 60 per cent rather than 70 per cent, and that would still have required him to pay £55 for a £100 bet on *North*. So, again, he would have been sure to lose £10. And indeed, unless he retains exactly the same credences between Monday and Tuesday, there will always be a pair of bets: one offered on Monday that his Monday credences require him to accept, and one offered on Tuesday that his Tuesday credences require him to accept, that, taken together, will lose him money at all epistemically possible worlds. So, if vulnerability to a diachronic Dutch Book is sufficient for irrationality, then Conrad is irrational, but so is anyone who ever changes any of their credences. And surely that can't be right.

## 4.2 Moderate Dutch Strategies and Rule Conditionalisation

This suggests that the existence of a diachronic Dutch Book against your actual updating behaviour is not sufficient for irrationality. But why not? One natural answer is this. Come Tuesday, we and Conrad will both know that he in fact learned *England* (which we'll call $E$) and updated his credence in *North* (which we'll abbreviate $N$) on the basis of that. But, on Monday, it is still open, at

---

[11]  We write $c(A|E)$ for your credence in $A$ conditional on $E$. When $c(E) > 0$, we define it as follows: $c(A|E) = \frac{c(A \ \& \ E)}{c(E)}$.

least from Conrad's own point of view, whether he will learn $E$ or not. Were he instead to learn $\overline{E}$, he might well have updated his credences in a different way. Let's suppose that, if he learns $E$, he'll become 70 per cent confident in $N$, and if he learns $\overline{E}$, he'll become 20 per cent confident in $N$. Then his Monday credences require him to sell a £100 bet on $N$ for £45, and his Tuesday credences should he learn $E$ require him to buy a £100 bet on $N$ for £55, thereby losing him money whether or not $N$ is true. But his Tuesday credences should he learn $\overline{E}$ do not require him to buy a £100 bet on $N$ for £55. The most they permit him to pay for that bet is £20, and that won't lose him money when combined with his Monday bet. So, while this pair of bets is guaranteed to lose him money if he enters into both, and while his prior credences and his posteriors if he learns $E$ require him to enter into both, his priors and his posteriors if he instead learns $\overline{E}$ do not.

This suggests the following sort of argument against someone who will update by something other than conditionalising in the face of certain evidence they might acquire. Suppose $c$ is your credence function at time $t$ – it is defined on $\mathcal{F}$. There's some proposition $E$ in $\mathcal{F}$ that you might learn as evidence between an earlier time $t$ and a later time $t'$. And you'll learn $E$ just in case it's true. And suppose $c(E) > 0$. If you learn $E$ – that is, if $E$ is true – you'll adopt credence function $c'$. If you don't learn $E$ – that is, if $E$ is false – we don't know how you'll respond – perhaps it isn't determined. Then we'll say that *there is a moderate Dutch Strategy against you* if there are sets of bets $B$, $B'_E$, $B'_{\overline{E}}$ such that

   (i)   $c$ requires you to accept each of the bets in $B$,
   (ii)  $c'$ requires you to accept each of the bets in $B'_E$,
   (iii) *any* credence function requires you to accept each of the bets in $B'_{\overline{E}}$,
   (iv)  taken together, the bets in $B$ and $B'_E$ lose you money in all worlds at which $E$ is true, and
   (v)   taken together, the bets in $B$ and $B'_{\overline{E}}$ lose you money in all worlds at which $E$ is false.

This is a particular diachronic version of de Finetti's notion of coherence. We'll say that you're irrational if you're vulnerable to a moderate Dutch Strategy, provided there are alternatives that are not vulnerable in this way. Now, if we accept this, we can give an argument for updating by conditionalising. Here's how.[12]

---

[12] I owe the presentation to Briggs (2009). The general form of argument originated in a handout that David Lewis produced for a graduate seminar at Princeton in 1972. He later published it (Lewis, 1999).

Suppose $c'(A) = r'$, $c(A|E) = r$, and $c(E) = d > 0$. Then consider the following bets:

| | Bet 1 (at $t$) | Bet 2 (at $t$) | Bet 3 (at $t'$ iff $E$) | Bet 4 (at $t'$ iff $\overline{E}$) | Net gain |
|---|---|---|---|---|---|
| $AE$ | $(1-r)S$ | $(d-1)(r'-r)S$ | $(r'-1)S$ | $0$ | $d(r'-r)S$ |
| $\overline{A}E$ | $-rS$ | $(d-1)(r'-r)S$ | $r'S$ | $0$ | $d(r'-r)S$ |
| $\overline{E}$ | $0$ | $d(r'-r)S$ | $0$ | $0$ | $d(r'-r)S$ |

So $B = \{$Bet 1, Bet 2$\}$, $B'_E = \{$Bet 3$\}$, and $B'_{\overline{E}} = \{$Bet 4$\}$. Whatever $S$ is, (i) Bets 1 and 2 have an expected utility of 0 relative to $c$, which is your credence function at $t$, (ii) Bet 3 has an expected utility of 0 relative to $c'$, which you will adopt at $t'$ if $E$ is true, and (iii) Bet 4 has expected utility of 0 relative to any credence function. Thus, your credences rationally *permit* you to accept each bet, if and when it is offered. Now, if $r' > r$, then let $S < 0$, and if $r' < r$, then let $S > 0$. Then, whether or not $E$ is true, the bets you'll accept, taken together, gain you $d(r' - r)S$ for sure, and that is negative. What's more, we can find a number $\varepsilon$ for which the following holds: if we add $\varepsilon$ to the utility of each outcome of each bet, then (i) Bets 1 and 2 have positive expected utility relative to $c$, and thus your credences at $t$ rationally *require* you to accept each bet, (ii) Bet 3 has positive expected utility relative to $c'$, and thus your credences at $t'$ if $E$ is true and you therefore learn $E$ rationally *require* you to accept it, (iii) Bet 4 has positive expected utility relative to any credence function, and (iv) whether $E$ is true or not, these bets taken together gain you $d(r' - r)S + 3\varepsilon$ for sure, and this is negative. Indeed, any $0 < \varepsilon < \frac{d(r-r')}{3}S$ will do the trick. Thus, if you will update in one particular way if you learn $E$ and if that way is not conditionalisation, then there is a moderate Dutch Strategy against you.

Now notice that this argument is directed against someone who will update by something other than conditionalisation on certain evidence she might receive. Thus, at least on the face of it, it is not directed against Conrad's *actual updating behaviour*, but rather against his *dispositions to update in different ways depending on the evidence he receives* – what we might call his *updating rule*. That is, the object of criticism against which the Dutch Strategy argument is posed is Conrad's updating rule. One way to see this is to ask what would happen if Conrad instead learned $\overline{E}$ and updated on $\overline{E}$ by conditionalising on it. Then, even though his actual updating behaviour would have been in line with Actual Conditionalisation, he would nonetheless still have been vulnerable to a Dutch Strategy because he would have strayed from conditionalisation

had he learned $E$ instead. This shows that Dutch Strategy arguments target irreducibly modal features of an agent – that is, they target rules or dispositions, not
actual behaviour. We will see this again below. Thus, we might take the Dutch
Strategy argument to establish the following law, at least in the first instance:

> **Rule Conditionalisation** If
>
> (i)   $c$ is your credence function at $t$,
> (ii)  if $E$ is the strongest evidence you obtain between $t$ and $t'$, then you will
>       adopt $c'$ as your credence function at $t'$,
> (iii) you will learn $E$ iff $E$ is true,
> (iv)  $c(E) > 0$,
>
> then it ought to be that
>
> (v)   $c'(A) = c(A|E)$, for all $A$.

The crucial difference between Actual and Rule Conditionalisation lies in the
modal status of the second clause. Whereas Actual Conditionalisation targets
what you *actually have* done, Rule Conditionalisation targets what you *will* do.

## 4.3 Strong Dutch Strategies and Generalised Rule Conditionalisation

Now, it might seem that we can salvage an argument for Actual Conditionalisation from Rule Conditionalisation. Conrad violates Actual Conditionalisation
because his unconditional credence in $N$ on Tuesday is 70 per cent while his
conditional credence in $N$ given $E$ on Monday is 60 per cent. But surely it
was then true on Monday that he *will* adopt a credence of 70 per cent in $N$ on
Tuesday *if he learns E*. That is, he violates Rule Conditionalisation as well.

But there is a problem with that reasoning. Suppose Conrad's credences don't
evolve deterministically. That is, suppose that, while on Tuesday it turns out that
he *in fact* responded to learning $E$ by raising his confidence in $N$ to 70 per cent,
he *might have* responded differently. For instance, suppose that there was some
possibility that he responded to the evidence $E$ by dropping his confidence to 50
per cent instead. Then the moderate Dutch strategy against Conrad described
above has a hole. It tells us what to do if he learns $E$ *and responds by becoming
70 per cent confident in N*. And it tells us what to do if he doesn't learn $E$. But
it says nothing about what to do if he learns $E$ *and he drops his confidence in
N to 50 per cent*. And indeed it turns out that it isn't always possible to fill that
gap.

So, what the standard Dutch Strategy argument sketched above shows is that
there is always a Dutch strategy against someone with a *deterministic* updating

rule that would make them stray from conditionalising in some cases. Now, it turns out that, for certain nondeterministic updating rules, we can create Dutch Strategies against them too. But not all of them. Indeed, there are nondeterministic ways to update your credences that *always* lead you to not conditionalise, but for which there is no strategy for creating a Dutch Book against you.

We represent an updating rule as a pair:

$$R = (\mathcal{E} = \{E_1, \ldots, E_m\}, \mathcal{C} = \{C_1, \ldots, C_m\}),$$

where

- $E_1, \ldots, E_m$ are the possible propositions you might learn as evidence, and
- for each $E_j$ in $\mathcal{E}$, $C_j$ is the set of credence functions you might adopt in response to learning $E_j$.

We assume that $\mathcal{E}$ is a partition. Thus, $R$ is deterministic iff $C_j$ is a singleton $\{c_j\}$ for all $1 \leq j \leq m$.

Now, let's suppose Conrad has the nondeterministic updating rule depicted in Figure 4.1. Thus, it is $R = (\mathcal{E}, \mathcal{C})$, where

- $\mathcal{E} = \{E, \overline{E}\}$,
- $C_E = \{c_1, c_2\}$, and
- $C_{\overline{E}} = \{c_3, c_4\}$.

Then, whatever happens, Conrad will not update by conditionalising on his evidence. But, there is no Dutch strategy against him. To see this, let me introduce a property of updating rules. Suppose $E_j$ is in $\mathcal{E}$ and $c'$ is in $C_j$. Then let $R^j_{c'}$ be the proposition that says that you adopt $c'$ as your posterior in response to receiving evidence $E_j$. Then we say:

- a rule $R$ is *a weak super-conditionalising rule* for a prior credence function $c$ if there is an extension of the prior $c$, which is defined on $\mathcal{F}$, to a credence function $c^{\dagger}$, which is defined on $\mathcal{F}^{\dagger} = (\mathcal{F} \cup \{R^j_{c'} : E_j \in \mathcal{E} \;\&\; c' \in C_j\})^*$, such that, for each $E_j$ in $\mathcal{E}$ and $c'$ in $C_j$, if $c^{\dagger}(R^j_{c'}) > 0$, then for all $A$ in $\mathcal{F}$,[13]

$$c^{\dagger}(A|R^j_{c'}) = c'(A)$$

That is, an updating rule is a weak super-conditionalising rule for a prior just in case it is possible to extend the prior so that it assigns credences to facts about the evidence you receive and the credences you adopt in response to it in such a way that, for any evidence you might receive and any possible

---

[13] Recall: $(\mathcal{F} \cup \{R^j_{c'} : E_j \in \mathcal{E} \;\&\; c' \in C_j\})^*$ is the smallest algebra that contains every proposition in $\mathcal{F}$ as well as each $R^j_{c'}$.

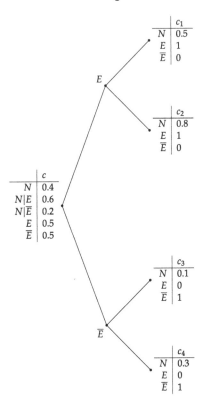

**Figure 4.1** Conrad's credences on Monday are given by $c$. On Tuesday, he learns $E$ or $\overline{E}$. If $E$, he adopts either $c_1$ or $c_2$, but it is not determined which. If $\overline{E}$, he adopts either $c_3$ or $c_4$, but it is not determined which.

response, if the extended prior assigns positive credence to that evidence and response, then that response is the result of conditionalising the extended prior on the proposition that you learned that evidence and responded in that way.

- a rule $R$ is *a strong super-conditionalising rule* for a prior credence function $c$ if there is an extension of the prior $c$, which is defined on $\mathcal{F}$, to a credence function $c^\dagger$, which is defined on $\mathcal{F}^\dagger = (\mathcal{F} \cup \{R^j_{c'} : E_j \in \mathcal{E} \,\&\, c' \in C_j\})^*$, such that, for each $E_j$ in $\mathcal{E}$ and $c'$ in $C_j$, (i) $c^\dagger(R^j_{c'}) > 0$, and (ii) for all $A$ in $\mathcal{F}$,[14]

$$c^\dagger(A|R^j_{c'}) = c'(A)$$

---

[14] Recall: $(\mathcal{F} \cup \{R^j_{c'} : E_j \in \mathcal{E} \,\&\, c' \in C_j\})^*$ is the smallest algebra that contains every proposition in $\mathcal{F}$ as well as each $R^j_{c'}$.

That is, an updating rule is a strong super-conditionalising rule for a prior just in case it is possible to extend the prior so that it assigns credences to facts about the evidence you receive and the credences you adopt in response to it in such a way that, for any evidence you might receive and any possible response, the extended prior does assign positive credence to that evidence and response, and the response is the result of conditionalising the extended prior on the proposition that you learned that evidence and responded in that way.

Notice that Conrad's updating rule, depicted in Figure 4.1, is a strong super-conditionalising rule for his Monday credence function $c$. To see this, for each $w$ in $\mathcal{W}_{\mathcal{F}}$, let

$$c^{\dagger}(w \,\&\, R^{E}_{c_1}) = \frac{1}{3}c_1(w)$$

$$c^{\dagger}(w \,\&\, R^{E}_{c_2}) = \frac{1}{6}c_2(w)$$

$$c^{\dagger}(w \,\&\, R^{\overline{E}}_{c_3}) = \frac{1}{4}c_3(w)$$

$$c^{\dagger}(w \,\&\, R^{\overline{E}}_{c_4}) = \frac{1}{4}c_4(w)$$

Before we state our central Dutch Strategy theorem, we provide some further diachronic versions of de Finetti's notion of coherence to supplement the notion of a *moderate Dutch Strategy* from page 38 above: We say that *there is a strong Dutch Strategy against you* if there are sets of bets $B$ and $B'$ such that

(i) $c$ requires you to accept each of the bets in $B$,
(ii) for each $E_j$ in $\mathcal{E}$ and each $c'$ in $C_j$, $c'$ requires you to accept each of the bets in $B'$, and
(iii) taken together, the bets in $B$ and $B'$ lose you money in all epistemically possible worlds.

Note that, in contrast with a moderate Dutch Strategy, the second set of bets $B'$ in a strong Dutch Strategy does not depend on the evidence you acquire.

We also say that *there is a weak Dutch Strategy against you* if there is a set of bets $B$ and, for each $E_j$ in $\mathcal{E}$ and each $c'$ in $C_j$, a set of bets $B'_{c'}$, such that

(i) $c$ permits you to accept each of the bets in $B$,
(ii) for each $E_j$ in $\mathcal{E}$ and each $c'$ in $C_j$, $c'$ permits you to accept each of the bets in $B'_{c'}$,
(iiia) for each $E_j$ in $\mathcal{E}$ and each $c'$ in $C_j$, the bets in $B$ and $B'_{c'}$, taken together, will not gain you money at any world in $E_j$ at which you adopt $c'$.

(iiib) for some $E_j$ in $\mathcal{E}$ and some $c'$ in $C_j$, the bets in $B$ and $B'_{c'}$, taken together, will lose you money at some world in $E_j$ at which you adopt $c'$.

Note that, in contrast with a moderate Dutch Strategy, the second set of bets $B'_{c'}$ in a weak Dutch Strategy depends not only on the evidence you acquire, but also on the posterior credence function you adopt in response to it.

**Theorem 4.1** *Suppose c is your prior and R is your updating rule. Then:*

(I) *If R is not a weak or strong super-conditionalising rule for c, there is at least a weak Dutch Strategy against you, and possibly a strong Dutch Strategy..*

(II) *If R is a strong super-conditionalising rule for c, there is no weak (nor strong) Dutch Strategy against you.*

(III) *If R is a weak super-conditionalising rule for c, there is no strong Dutch Strategy against you.*

We sketch the proof in Section 8.

This gives us the following Dutch Strategy argument for a generalised version of Rule Conditionalisation.

(DB1)  If you have credence $p$ in proposition $A$, then

(i) you are rationally permitted to pay £$x$ for a £$S$ bet on $A$ for any $x \leq pS$,

(ii) you are rationally required to pay £$x$ for a £$S$ bet on $A$ for any $x < pS$.

(DSC2) **Theorem 4.1**

(DSC3) If your prior and updating rule are vulnerable to a weak or strong Dutch Strategy, and there are alternative priors and updating rules that are not vulnerable to a weak or strong Dutch Strategy, then you are irrational.

Therefore,

(DSCC) **Generalised Rule Conditionalisation** It ought to be that your updating rule $R$ is at least a weak super-conditionalising rule for your prior $c$.

Now, if your updating rule is deterministic, then it is a weak super-conditionalising for your prior iff it is conditionalising for it. A deterministic updating rule is vulnerable to a strong Dutch Strategy if it is not a conditionalising rule, and not vulnerable even to a weak Dutch Strategy if it is. Thus, we have an extra argument for Rule Conditionalisation. In some ways it strengthens the standard argument presented above, for it shows that the same set of bets can be offered at $t'$ regardless of what credence function you end up having.

But in some ways it weakens that argument, for it relies on the assumption that there is a finite set of possible future credence functions you might adopt at $t'$.

Now, of course, you might object that updating other than by a deterministic rule is irrational:[15] your evidence, together with your prior credences, should determine your new credences; there should not be many different possible ways you might respond to the same piece of evidence. This may be true, and if we supplement the Dutch Strategy argument with this assumption, we obtain an argument for Plan Conditionalisation and possibly also Actual Conditionalisation. But note that the argument no longer establishes the irrationality of failing to conditionalise by appealing only to the decisions that the credences endorsed by such updating rules lead you to make—that is, it is no longer purely pragmatic.[16]

## 4.4 Dutch Strategies and the Reflection Principle

In this section, we turn to an objection against such Dutch Strategy arguments for various forms of conditionalisation. As van Fraassen (1984) pointed out, we can give a seemingly analogous argument for his Reflection Principle. Given a possible posterior credence function $c'$, let $R_{c'}$ say that $c'$ is your posterior.

**Reflection Principle** If $c(R_{c'}) > 0$, then $c(A|R_{c'}) = c'(A)$.

Essentially, the Dutch Strategy argument for the Reflection Principle is obtained from the Dutch Strategy argument for Plan Conditionalisation by replacing the evidence $E$ in the latter with the proposition $R_{c'}$.

However, the Reflection Principle demands a level of deference to your future credences that is sometimes simply not rationally permitted, let alone required. For instance, if Conrad knows that between Monday and Tuesday he will take a drug that makes him irrationally under-confident in southerly landfalls for summer storms and over-confident in northerly ones, his confidence on Monday that Storm Saoirse will land north of the Lake District conditional on him being 90 per cent confident on Tuesday that it will, for instance, should not be 90 per cent – it should be less than that.[17] Thus, from the falsity of the Reflection Principle, many (though not van Fraassen himself) infer that the Dutch Strategy argument in its favour must be invalid; and from that they infer

---

[15] The remarks that David Lewis makes towards the end of Lewis (1999) suggest that he might think this.

[16] For further discussion of this point, see Pettigrew (2019b).

[17] For other counterexamples, see Talbott (1991), Christensen (1991), and Arntzenius (2003). For an overview, see Briggs (2009, 64–66).

that all such arguments are invalid, and thus they cast doubt on the particular Dutch Strategy argument for conditionalising.

In the remainder of this section, we'll consider R. A. Briggs' attempt to show that the Dutch Strategy argument for Rule Conditionalisation works, while the argument for the Reflection Principle doesn't. Then we'll see Anna Mahtani's response, where she argues that, while Briggs' argument shows that there is no Dutch Strategy argument for the Reflection Principle, there is one for a weaker principle that is implausible for the same reasons that the original principle is implausible. If Mahtani is right, as I will argue she is, the apparent objection against Dutch Strategy arguments and thus against the Dutch Strategy argument for Rule Conditionalisation is revived. I then argue that the apparent Dutch Strategy argument for Reflection does not establish that principle, and thus that the validity of Dutch Strategy arguments is unimpeached.

Briggs (2009) argues that, contrary to appearances, there is a disanalogy between the Dutch Strategy argument for Rule Conditionalisation and the Dutch Strategy argument for the Reflection Principle. This allows us to reject the argument for the Reflection Principle as invalid, while retaining the argument for conditionalising as valid. To see Briggs' point, let's place the two arguments side by side:

- First, the Dutch Strategy argument for Rule Conditionalisation. Suppose $c$ is my prior, suppose $c(E) > 0$, and suppose that, if I learn $E$, I will adopt $c'$ as my posterior, where $c'(A) \neq c(A|E)$, for some $A$. Then there are sets of bets $B$, $B'_E$, and $B'_{\overline{E}}$ such that
  (i) $c$ requires me to accept $B$,
  (ii) $c'$ requires me to accept $B'_E$, and
  (iii) any credence function requires me to accept $B'_{\overline{E}}$,
  and
  (a) taken together, $B$ and $B'_E$ will lose me money in all worlds at which $E$ is true, and
  (b) taken together, $B$ and $B'_{\overline{E}}$ will lose me money in all worlds at which $E$ is false.

- Second, the Dutch Strategy argument for the Reflection Principle. Suppose $c$ is my prior, suppose $c(R_{c'}) > 0$, and suppose that $c(A|R_{c'}) \neq c'(A)$, for some $A$. Then there are bets $B$, $B'_{R_{c'}}$, and $B'_{\overline{R_{c'}}}$ such that
  (i) $c$ requires me to accept $B$,
  (ii) $c'$ requires me to accept $B'_{R_{c'}}$, and
  (iii) any credence function requires me to accept $B'_{\overline{R_{c'}}}$
  and
  (a) taken together, the bets in $B$ and $B_{R_{c'}}$ will lose me money in all worlds at which my posterior is indeed $c'$;

(b) taken together, $B$ and $B'_{\overline{R_{c'}}}$ will lose me money in all worlds at which it is not.

Now, notice a gap in the formulation of the Dutch Strategy argument for Rule Conditionalisation. Suppose it is epistemically possible that $E$ is true but I don't learn $E$. Then something other than $c'$ might be my posterior, even though it is epistemically possible for me that $E$ is true. In that case, the strategy would offer me the bets in $B'_{\overline{E}}$. But there is no guarantee that, when taken together with the bets in $B$, this loses me money at this epistemically possible world at which $E$ is true. We solve this, of course, by stating explicitly in Rule Conditionalisation, that you learn $E$ iff $E$ is true. That is, Rule Conditionalisation only applies in such cases, and thus the Dutch Strategy argument establishes it.

Now, Briggs notes that there is a similar gap in the formulation of the Dutch Strategy argument for the Reflection Principle. Suppose it is epistemically possible that $c'$ is my posterior but my evidence at the time I have it doesn't entail that. Then $c'$ might be my posterior, even though it is epistemically possible for me that $R_{c'}$ is false – just as in the previous case, something other than $c'$ might be my posterior, even those it is epistemically possible for me that $E$ is true. In that case, the strategy would offer me the bets in $B'_{R_{c'}}$. But there is no guarantee that, when taken together with the bets in $B$, this loses me money at the epistemically possible world at which $R_{c'}$ is false. This is the problem that Briggs identifies for the Dutch Strategy argument for the Reflection Principle.

However, while Briggs' observation successfully blocks the argument for the Reflection Principle in its strong, general form, Anna Mahtani (2012) points out that it does not block a Dutch Strategy argument for a weaker, more specific version of the law:

> **Weak Reflection Principle** If at $t'$ you will have conclusive evidence concerning what your credence function is, then if $c(R_{c'}) > 0$, then $c(A|R_{c'}) = c'(A)$.

This essentially restricts the Reflection Principle in the same way that we restricted Rule Conditionalisation in order to fix the analogous problem there. In the case of Rule Conditionalisation, we include the assumption that you learn $E$ iff $E$ is true; in the case of Mahtani's Weak Reflection Principle, we include the assumption that you have posterior $c'$ iff your evidence entails that you do. So now the Dutch Strategy argument establishes the Weak Reflection Principle.

Now, the Weak Reflection Principle may be weaker than the original version, but it is still very implausible. Conrad should not be any more inclined to defer to the credences he will end up having after taking the drug just because his evidence at the later time will tell him exactly what his credences are at that time. Thus, the objection to the Dutch Strategy argument for Rule Conditionalisation remains intact.

How, then, should we respond to this objection? The first thing to note is that, in a sense, the Dutch Strategy argument for Reflection does not actually target Reflection. After all, unlike the versions of conditionalisation we have been considering, Reflection is a synchronic law. It says something about how your credences should be at $t$. It says nothing about how your credences at $t$ should relate to your credences at $t'$, only how your credences at $t$ *about your credences at $t'$* should relate to your other credences at $t$. But the Dutch Strategy argument involves bets that your credences at $t$ require you to accept, as well as bets that your credences at $t'$ require you to accept. You can violate Reflection whilst having probabilistic credences. So the Converse Dutch Book Theorem shows that there is no synchronic Dutch Book argument against your credences – that is, there is no set of bets that $c$ alone requires you to accept that will lose you money at all epistemically possible worlds.

So what's going on? The key fact is this: if you violate Reflection, and you have a deterministic updating rule, then that updating rule cannot possibly be a conditionalising rule. After all, suppose $c(R_{c'}) > 0$ and $c(A|R_{c'}) \neq c'(A)$ and you learn $R_{c'}$ and nothing more. Then, since you learn $R_{c'}$, it must be true and thus your new credence function must be $c'$. But your violation of Reflection ensures that $c'$ is not the result of conditionalising on your evidence $R_{c'}$. So the Dutch Strategy argument for Reflection does not target Reflection itself; rather, it targets the updating rule you are forced to adopt because you violate Reflection, if you decide to adopt an updating rule at all.

Consider the analogous case.[18] Suppose you think that it is irrational to have a set of beliefs that can't possibly all be true together. Now, suppose you have the following second-order belief: you believe that you believe a contradiction, such as $A \& \overline{A}$. Then that second-order belief might be true. So, by your standards, it is not irrational in this way. However, suppose we now consider what your attitude to $A \& \overline{A}$ is. Whatever attitude you have, you are guaranteed to have a false belief: if you do believe the contradiction, your second-order belief is true, but your first-order belief is false; if you don't believe the contradiction, then the second-order belief itself is false. In this case, we might say that the belief itself is not irrational—it might be true, and it might be supported by your evidence. But its presence forces your total doxastic state to be irrational.

The same thing is going on in the case of Reflection. Just as you think that it is irrational to have beliefs that cannot all be true, you also think it is irrational to have credences that require you to enter into bets that lose you money for

---

[18] Briggs (2009, 78–83) also considers this analogy, but draws a different moral from it.

sure. And just as the single second-order belief that you believe a contradiction is possibly true, so, by the Converse Dutch Book Theorem, a probabilistic credence function that violates Reflection doesn't require you to accept any bets that will lose you money for sure. However, just as the single belief that you believe a contradiction forces you to have an attitude to the contradiction (either believing it or not) that ensures that your total doxastic state (first- and second-order beliefs together) includes a false belief, so your violation of Reflection forces you to adopt an updating rule that is vulnerable to a Dutch Strategy. For this reason, we can allow that you are irrational if you are vulnerable to a Dutch Strategy without rendering violations of Reflection irrational, just as we can allow that you are irrational if you have beliefs that are guaranteed to include some falsehoods without rendering the second-order belief that you believe a contradiction irrational. Both force you to adopt some other sort of doxastic state—a first-order belief or an updating rule—that is irrational. But they are not themselves irrational. This saves the Dutch Strategy argument for Rule Conditionalisation.

In sum: If you have a deterministic updating rule, the Dutch Strategy argument for Rule Conditionalisation establishes that you must update by conditionalisation on your prior. But there is no Dutch Strategy argument that shows you are irrational if you don't have a deterministic updating rule. To establish that, we must appeal to considerations outside the scope of betting arguments. If you don't have a deterministic updating rule, then the Dutch Strategy argument for Generalised Rule Conditionalisation establishes that your updating rule must be a super-conditionalising rule for your prior.

## 5 The Choices Credences Rationally Require

At the end of Section 3, we formulated precise versions of the traditional Dutch Book arguments for Probabilism, Regularity, and the Principal Principle. These versions of the arguments avoid many of the initial objections that such arguments face. In this section, we consider two objections to these arguments that challenge the first premise, namely, (DB1). In the next section, we consider a further objection, which targets the third premises, (DB3FP), (DB3FPR), and so on.

### 5.1 The Expected Utility Objection

The first objection is due to Brian Hedden (2013).[19] Recall: (DB1) says that, for any stake $S$, if you have credence $p$ in $A$, you're permitted to pay any price

---

[19] Wroński and Godziszewski (2017) raise a closely related worry.

up to and including $£pS$ for a $£S$ bet on $A$, and you're required to pay anything up to but not including that price. Now, when we introduced that normative claim, we justified it as follows: if you have credence 0 in a proposition, you're permitted to buy a $£S$ bet on it for £0, but no more; if you have credence 1, you're permitted to buy that bet for anything up to $£S$; and, if you have any credence in between, the maximum price you're permitted to pay rises linearly with the credence. And the same applies for the prices you're required to pay. However, Hedden claims, we shouldn't have to appeal to such a motivation. After all, we already have a general theory of how credences and utilities should guide action, and thus a theory of what bets your credences should require you to accept. But unfortunately, this general theory does not in fact support (DB1). Indeed, in some situations, its demands are incompatible with the demands of (DB1). And, what's more, if we replace (DB1) with this general theory as the first premise of the Dutch Book argument, the argument is rendered invalid and therefore fails. Or so says Hedden. In this section, I'll spell out his objection in more detail and I'll suggest two ways we might rewrite the Dutch Book argument to avoid it.

The general theory that Hedden has in mind is *expected utility theory*, which we met in Section 3.7 above. The central law of expected utility theory is (MEU), which says that, when faced with a decision problem, you are rationally required to choose an act from those available that has maximal expected utility relative to your credence and utility functions.

That is, if $c$ is your credence function and $u$ your utility function, you are rationally required to choose an act $a$ such that, for all available $b$,

$$\mathrm{EU}_{c,u}(b) = \sum_{o \in \mathcal{O}} c(b \Rightarrow o)u(o) \leq \sum_{o \in \mathcal{O}} c(a \Rightarrow o)u(o) = \mathrm{EU}_{c,u}(a)$$

Now, (MEU) is typically applied when your credences are probabilistic.[20] When they are, (MEU) and (DB1) can't conflict; when they aren't, they can. Consider Cináed. He is 60 per cent confident it will rain and 20 per cent confident it won't. According to (DB1), Cináed is rationally required to sell a £100 bet on rain for £65. But the expected utility of this bet for him is

$$0.6 \times (-100 + 65) + 0.2 \times (-0 + 65) = -8$$

---

[20] In what follow, I will grant Hedden's assumption that (MEU) is a viable decision theory. But it is worth saying that, if $c$ is non-probabilistic, and if utility is specified only up to positive affine transformation, the ordering of two options $a$ and $b$ by their expected utility relative to $c$ might not be well defined. That is, there might be utility functions $u_1$, $u_2$ such that (i) $u_1 = \alpha u_2 + \beta$ for $\alpha, \beta$ with $\alpha > 0$ and (ii) there are acts $a, b$ such that $\mathrm{EU}_{c,u_1}(b) < \mathrm{EU}_{c,u_1}(a)$ but $\mathrm{EU}_{c,u_2}(a) < \mathrm{EU}_{c,u_2}(b)$. For instance, in the example on page 52 below, notice that we can reverse the expected utility ordering of $a$ and $b$ by subtracting 70 from each of the utilities.

$$\{w_1, w_2, w_3\}$$
$$0.9 \parallel 1$$

| $\{w_1, w_2\}$ | $\{w_1, w_3\}$ | $\{w_2, w_3\}$ |
|---|---|---|
| $0.8 \parallel 1$ | $0.8 \parallel 1$ | $0 \parallel 0$ |

| $\{w_1\}$ | $\{w_2\}$ | $\{w_3\}$ |
|---|---|---|
| $0.7 \parallel 1$ | $0 \parallel 0$ | $0 \parallel 0$ |

$$\emptyset$$
$$0 \parallel 0$$

**Figure 5.1** Let $c$ be the non-probabilistic credence function that assigns the credences on the left of the divider ($\parallel$), while $c^\dagger$ is the probabilistic credence function that assigns the credences on the right.

assuming, as usual, that his utility is linear in money. That is, it has lower expected utility than refusing to sell the bet, since his expected utility for doing that is

$$0.6 \times 0 + 0.2 \times 0 = 0$$

While (DB1) says you must sell that bet for that price, (MEU) says you must not.[21] So, (DB1) and (MEU) are incompatible. Hedden claims that we should favour the latter. Let's consider how the Dutch Book argument fares in its presence.

As Hedden shows, if we take (MEU) to determine the bets that your credences require you to accept, then there are non-probabilistic credences that are not dutchbookable, and so the argument for Probabilism fails. In Hedden's illustrative example, your non-probabilistic credence function $c$ is defined on the algebra built up from three possible worlds, $w_1$, $w_2$, and $w_3$. It is illustrated in Figure 5.1 along with another credence function $c^\dagger$, which is probabilistic. The crucial fact is this: if (MEU) is correct, $c$ requires you to accept a bet iff $c^\dagger$ does. Since $c^\dagger$ is probabilistic, we can appeal to the Converse Dutch Book theorem: $c^\dagger$ is not dutchbookable, and so neither is $c$. Thus, if we replace (DB1) with (MEU), the Dutch Book argument seems to fail.

How might we respond to Hedden's objection? I see two possibilities.[22]

First, we might argue that, while (MEU) is a plausible decision theory, so is (DB1). Hedden notes that they conflict, assumes that we must pick one or the

---

[21] It's worth noting that some philosophers motivate (DB1) by appealing to expected utility theory. So it might seem surprising that (MEU) and (DB1) can conflict. Suppose you have credence $p$ in $A$. Then, they say, you should choose as if your credence in $\overline{A}$ were $1 - p$. And, if you do this, then (MEU) entails (DB1) because $x < pS$ iff $p(S - x) + (1 - p)(-x) > 0$. Hedden's point is that you should not choose as if your credence in $\overline{A}$ were $1 - p$. You should choose using your actual credence in $\overline{A}$. And, when you do that, (MEU) and (DB1) can conflict.

[22] The following builds on Pettigrew (2019a).

other, and favours (MEU). But perhaps we should say instead that, faced with the possibility of entering into a bet on a proposition, we are rationally permitted to do anything that (DB1) considers rationally permissible *and* we are rationally permitted to do anything that (MEU) considers rationally permissible. If that's the case, we can rehabilitate the Dutch Book argument by noting that, if you have a non-probabilistic credence function, then you may not be rationally *required* to accept each of the bets in a Dutch Book, but you are rationally *permitted* to do so. That is, you are finitely weakly fully exploitable. And while this is perhaps less irrational than being finitely strongly fully exploitable, it is nonetheless still irrational.[23]

Second, we might note that entering into each of a series of bets that, taken together, lose you money is not the only way in which *c* might lead you to make a bad choice. Consider the decision problem with only two available acts, *a* and *b*, whose outcomes, along with the utilities of those outcomes, are set out here:

|   | $w_1$ | $w_2$ | $w_3$ |
|---|-------|-------|-------|
| *a* | 78 ($o_1$) | 77 ($o_2$) | 77 ($o_2$) |
| *b* | 74 ($o_1'$) | 74 ($o_1'$) | 75 ($o_2'$) |

Then notice first that *a* dominates *b* – that is, the utility of *a* is higher than the utility of *b* in every possible state of the world. But:

$$EU_{c,u}(a) = c(a \Rightarrow o_1)u(o_1) + c(a \Rightarrow o_2)u(o_2)$$
$$= (0.7 \times 78) + (0 \times 77) = 54.6$$

$$EU_{c,u}(b) = c(b \Rightarrow o_1')u(o_1') + c(b \Rightarrow o_2')u(o_2')$$
$$= (0.8 \times 74) + (0 \times 75) = 59.2$$

So (MEU) demands that you choose *b*. In sum: if (MEU) is true, then *c* is not dutchbookable, but there is a decision problem (namely, choose *a* vs. choose *b*) in which *c* requires you to choose a dominated act (namely, *b*).

Now, this raises the question: For which credence functions does (MEU) lead you to choose a dominated act, and for which does it not? To state the answer, we need a weakened version of Normalisation: Normalisation says that your credence in a necessarily true proposition $\top$ should be maximal (i.e. 1);

---

[23] As we will see in Section 7.3 below, in the framework of imprecise credences, Katie Steele, Seamus Bradley, and Sarah Moss all argue that weakly exploitable credences are not irrational (Bradley & Steele, 2014; Moss, 2015). However, their reasons for doing so only apply in the context of imprecise credences.

Bounded Normalisation says that there is some $0 < M \leq 1$ such that your credence in a necessarily true proposition $\top$ should be $M$. It is the conjunction of Finite Additivity and Bounded Normalisation that we can establish if we replace the first premise of the Dutch Book argument with (MEU) – we call it *Bounded Probabilism*. Here's the relevant theorem:

**Theorem 5.1**

(I) *If c violates Bounded Probabilism, there are acts a, b, such that*
   (i) *a dominates b, and*
   (ii) $\mathrm{EU}_{c,u}(a) < \mathrm{EU}_{c,u}(b)$.
(II) *If c satisfies Bounded Probabilism, then, for all acts a, b, if*
   (i) *a dominates b,*
  *then*
   (ii) $\mathrm{EU}_{c,u}(a) > \mathrm{EU}_{c,u}(b)$.

We sketch the proof in Section 8. Thus, with (MEU) instead of (DB1), you can give a pragmatic argument for a norm that lies very close to Probabilism. On its own, this argument cannot say what is wrong with someone who is less than maximally confident in necessary truths, but it does establish the other requirements that Probabilism imposes. To see just how close Bounded Probabilism lies to Probabilism, note that $c$ satisfies Bounded Probabilism iff there is a credence function $c^\dagger$ that satisfies Probabilism and $0 < M \leq 1$ such that $c(-) = M \times c^\dagger(-)$.

So, on its own, (MEU) can deliver us very close to Probabilism. But it cannot establish full Normalisation. However, I think we can do better. Consider Lowri and Hyana. They both have minimal confidence (i.e. 0) that it won't rain tomorrow. But Lowri has credence 0.01 that it will rain, while Hyana has credence 0.99 that it will. If we permit only actions that maximise expected utility, then Lowri and Hyana are required to pay exactly the same prices for bets on rain – that is, Lowri will be required to pay a price exactly when Hyana is. After all, if £$S$ is the payoff when it rains, £0 is the payoff when it doesn't, and £$x$ is a proposed price, then $0.01 \times (S-x) + 0 \times (0-x) \geq 0$ iff $0.99 \times (S-x) + 0 \times (0-x) \geq 0$ iff $S \geq x$. So, according to (MEU), Lowri and Hyana are rationally required to pay anything up to the stake of the bet for such a bet. But this is surely wrong. It is surely at least permissible for Lowri to refuse to pay a price that Hyana accepts. It is surely permissible for Hyana to pay £99 for a £100 bet on rain, and permissible for Lowri to refuse to pay anything more than £1 for such a bet, in line with (DB1). Suppose Lowri were offered such a bet for the price of £99, and suppose she then defended her refusal to pay that price saying, 'Well,

I only think it's 1 per cent likely to rain, so I don't want to risk such a great loss with so little possible gain when I think the gain is so unlikely'. Then surely we would accept that as a rational defence.

In response to this, defenders of (MEU) might concede that (DB1) is sometimes the correct norm of action when you have non-probabilistic credences, but only in very specific cases, namely, those in which you have a positive credence in a proposition, minimal credence (i.e. 0) in its negation, and you are considering the price you might pay for a bet on that proposition. In all other cases – that is, in any case in which your credences in the proposition and its negation are both positive, or in which you are considering an action other than a bet on a proposition – you should use (MEU). But, of course, it is precisely by applying (DB1) to such a case that we can produce a Dutch Book against someone with $c(\bot) = 0$ and $c(\top) = M < 1$ – we simply offer to pay them something between £$M \times 100$ and £100 for a £100 bet on $\top$; this is then guaranteed to lose them money.

Thus, we end up with a disjunctive norm, which we'll call (MEU*). It says: apply (DB1) when you are deciding whether or not to accept a bet on a proposition and when you assign zero credence to the proposition or its negation and positive credence to the other; and apply (MEU) in all other cases.

Based on this, we end up with a disjunctive pragmatic argument for Probabilism. I call it the *Bookless Pragmatic Argument for Probabilism*:

- First, if $c(\bot) = 0$ and $c(\top) = M < 1$, then (DB1) applies and we can produce a Dutch Book against you in the traditional way. If we then assume that $c$ is irrational if it requires you to accept a bet that loses you money for sure, then $c$ is irrational.
- Second, if $c$ violates Probabilism in any other way, then it violates Bounded Probabilism. In that case, (MEU) applies and Theorem 5.1 ensures that there is a dominated act that your credences require you to choose. If we then assume that $c$ is irrational if it requires you to choose a dominated act, then $c$ is irrational.

Putting these together, in the presence of (MEU*), if a credence function violates Probabilism, it is singly strongly fully exploitable, and thus irrational.[24]

---

[24] It's worth noting that this does not furnish us with an argument for Full Probabilism.

## 5.2 The Risk-Sensitivity Objection

In the previous section, we saw Brian Hedden argue that we should replace (DB1) with a norm governing betting decisions that he takes to be closer to the central norm of decision theory for probabilistic agents, namely, maximise subjective expected utility (MEU). But some decision theorists have argued that expected utility theory is the wrong decision theory even for probabilistic individuals. Their objection is that expected utility theory does not accommodate rational agents who are either risk-averse or risk-seeking (Allais, 1953; Buchak, 2013).

Suppose I prefer £50 for sure to a fair coin toss that will give me £100 if the coin lands heads and £20 if it lands tails. As we noted in Section 3.3, this might be because money has diminishing marginal utility for me, so that my utility for £50 is greater than the average of my utility for £100 and my utility for £20 (which is £60). But suppose money doesn't have diminishing marginal utility for me. Suppose my utility is linear in money. Then it still seems I am rationally permitted to prefer the sure thing to the coin toss. After all, while the coin toss gives a possibility of much more utility than the sure thing, it also gives the possibility of much less. And, if I am risk averse, I give the worst outcome greater weight in my deliberations than my credences suggest, and I give the best outcome less weight than my credences suggest. So, I might prefer the sure thing to the gamble, even though my utility is linear in money. And, of course, this is contrary to the norm of expected utility theory, which underpins (MEU).

Now, there are many non-expected utility theories, each of which claims to accommodate certain violations of expected utility theory due to risk aversion.[25] I'll focus on Lara Buchak's *risk-weighted expected utility theory (REU)*, which is the most sophisticated and plausible of the available options (Buchak, 2013). According to REU theory, we represent individuals as having not only a credence function and a utility function, but also a risk function, which skews the credences that agents assign to the different outcomes of an action in order to give greater weight to certain of those outcomes than the credences alone would do in expected utility theory.

Here, I'll only explain how REU theory works for actions, such as bets, where there are just two outcomes – you win the bet or you lose it. And we'll consider only cases in which your credences are probabilistic. Consider an action $a$ with just two outcomes, $A$ and $\overline{A}$. If $A$, you receive utility $u'$; if $\overline{A}$, $u$. And

---

[25] Some take the violations they accommodate to be rationally permissible; others take them to be rationally forbidden. For the latter, the purpose of their theory is descriptive – they wish to accommodate observed choice behaviour but not to claim that it is rational.

suppose $u < u'$, so that $\overline{A}$ is the worst-case scenario and $A$ is the best-case scenario. Suppose $p$ is your credence in $A$ and $1-p$ is your credence in $\overline{A}$. Then the expected utility of $a$ can be written as the sum of the minimum utility you'll receive for sure (namely, $u$) and the extra utility you might receive (namely, $u' - u$) weighted by your credence you'll receive it (namely, $p$). That is,

$$\text{EU}_{p,u}(a) = u + p(u' - u)$$

The risk-weighted expected utility of $a$, on the other hand, is the sum of the minimum utility you'll receive for sure (namely, $u$) and the extra utility you might receive (namely, $u' - u$) weighted by your risk-transformed credence you'll receive it (namely, $r(p)$, where $r$ is your risk function):

$$\text{REU}_{p,u,r}(a) = u + r(p)(u' - u)$$

If $r(p) < p$, then you give less weight to the best-case scenario than expected utility theory demands, and in this case we say you're risk averse; if $r(p) > p$, you give more weight, and we say you're risk seeking; and if $r(p) = p$, then risk-weighted expected utility coincides with expected utility, and we say you're risk neutral. For instance, suppose $r(0.5) < 0.375$. Then, the risk-weighted expected utility of the coin toss is

$$
\begin{aligned}
\text{REU}_{p,u,r}(\textit{Coin Toss}) &= 20 + r(0.5)(100 - 20) \\
&< 20 + (0.375 \times 80) \\
&= 50 \\
&= \text{REU}_{p,u,r}(\textit{Sure Thing}) = 50
\end{aligned}
$$

So, an individual with that risk function and utility function should take the sure thing rather than the gamble – while the gamble has higher expected utility, it has lower risk-weighted expected utility.

So, if we follow Buchak, (DB1) from Section 3 and (MEU*) from Section 5 are both mistaken. Their demands are too strong. But, just as we replaced (DB1) with. (MEU*) and used that to mount an alternative pragmatic argument for Probabilism, so might we replace both with a risk-sensitive decision principle and use that to mount a different alternative pragmatic argument. In fact, as we will see, it is not clear whether this is possible.

One way to understand (DB1) is as follows: suppose you have credence $p$ in $A$. Then, if your credences were probabilistic, you'd have credence $1 - p$ in $\overline{A}$, and your expected utility for paying price £$x$ for a £$S$ bet on $A$ would be

$$p \times (-x + S) + (1 - p) \times (-x)$$

which is positive iff $x < pS$, and thus expected utility theory demands that you pay $£x$ for that bet iff $x < pS$ (see footnote 21). In the light of this, we might create a risk-sensitive version of (DB1) in a similar way (as above, we assume that your utility is linear in money):

(DB1$_r$) If you have credence $p$ in proposition $A$, then

(i) If $S > 0$, then $-x < -x + S$. As a result, $\overline{A}$ is the worst-case scenario. So, if your credence in $\overline{A}$ were $1-p$, your risk-weighted expected utility for paying price $£x$ for a $£S$ bet on $A$ would be $-x + r(p)S$, while your utility for refusing it would be 0. So paying the price would have greater risk-weighted expected utility than refusing the bet iff $-x + r(p)S > 0$ iff $x < r(p)S$.

So you are rationally permitted to pay $£x$ for a $£S$ bet on $A$ for any $x \leq r(p)S$, and rationally required for any $x < r(p)S$.

(ii) If $S < 0$, then $-x + S < -x$. As a result, $A$ is worst-case scenario. So, if your credence in $\overline{A}$ were $1 - p$, your risk-weighted utility for paying price $£x$ for a $£S$ bet on $A$ would $(-x + S) + r(1 - p)(-S)$, while your utility for refusing it would be 0. So paying the price would have greater risk-weighted expected utility than refusing the bet iff $(-x + S) + r(1 - p)(-S) > 0$ iff $x < (1 - r(1 - p))S$.

So you are rationally permitted to pay $£x$ for a $£S$ bet on $A$ for any $x \leq (1 - r(1 - p))S$, and rationally required for any $x < (1 - r(1 - p))S$.

The problem with this proposal is that it puts certain non-probabilistic credences beyond the reach of the Dutch Book argument. Consider, for instance, an individual whose credence in $A$ is 0.3, in $B$ is 0.4, but whose credence in their disjunction, $A \lor B$, is only 0.6. Then, if their risk function is the risk-averse $r(p) = p^2$, then it's possible to show that, for any stakes $S_1, S_2, S_3$, there are no prices $x_1, x_2, x_3$, such that risk-weighted expected utility permits that this individual pays $£x_i$ for a $£S_i$ bet on the relevant proposition, and such that the total price exceeds the total payout in any world. That is, such an agent is not finitely weakly fully exploitable when we adopt risk-weighted expected utility theory rather than expected utility theory. I have no response to offer to this.

## 6 The Alleged Irrationality of Being Exploitable

The objections we considered in the previous sections targeted the first premise of Dutch Book arguments, (DB1). In this section, we consider an objection to their third premises – (DB3N), (DB3FA), etc. – which infer irrationality from various kinds of exploitability.

## 6.1 Is It Irrational to Be Exploitable?

The problem is this: It is clear that it is irrational to make a series of decisions when there is an alternative series that is guaranteed to result in greater utility. It is irrational because, when you act, you are attempting to maximise your utility, but doing what you have done is guaranteed to be suboptimal as a means to that end – there is an alternative that you can know a priori would serve that end better. But it is much less clear why it is irrational to have credences that require you to make a dominated series of decisions when faced with a particular decision problem.

When you choose a dominated option, you are irrational because there's something else you could have done that is guaranteed to serve your ends better. But when you have credences that require you to choose a dominated option in a very specific situation, that alone doesn't tell us that there is anything else you could have done – any alternative credences you could have had – that would be guaranteed to serve your ends better.

Of course, there are credences you could have had that would not have required you to make the dominated choice. The Converse Dutch Book Theorem shows that, if your credences are probabilistic, then there is no dominated series of decision problems that your credences require you to make. So it's true that there's something else you could do that's guaranteed not to require you to make a dominated choice. But that only shows that there's some alternative that is better *in one particular way*. It doesn't show that it is better overall, nor that by being better in this way, you are guaranteed to receive better outcomes. After all, for all the Dutch Book or Converse Dutch Book Theorem tells you, it might be that your non-probabilistic credences lead you to choose badly when faced with the very particular Dutch Book decision problem, but lead you to choose extremely profitably when faced with many other decision problems. Furthermore, you might think it extremely unlikely that you will ever face the Dutch Book decision problem itself, or at least much more probable that you'll face other decision problems where your credences don't lead you to choose a dominated series of options; perhaps other decision problems where your credences will serve you well.

For all these reasons, the mere possibility of a finite series of decision problems from which your credences require you to choose a dominated series of options is not sufficient to show that your credences are irrational. To do this, we need to show that there are some alternative credences that are in some sense sure to serve you better as you face the decision problems you encounter in your life. Without an alternative that does better, pointing out a flaw in a mental state does not show that it is irrational, even if there are other mental states without the flaw – for those alternative mental states might have other strikes against them that the mental state in question does not have.

So our question is now: Is there any sense in which, when you have non-probabilistic credences, there are some alternative credences that are guaranteed to serve you better as a guide in your decision-making? Building on work by Mark Schervish (1989) and Ben Levinstein (2017), I'll argue that there is.

## 6.2 Is It Irrational to Have Inconsistent Preferences?

However, before we embark on this, I would like to consider a popular response to the objection presented in the previous section.[26] According to this response, it doesn't matter whether or not you will ever in fact face the series of decision problems in response to which your credences require you to make a dominated series of choices. Exploitable credences are not irrational because of any anticipated ill effects they will have for you if you hold them. Rather, what is irrational about such credences is that they generate inconsistent preferences, and inconsistent preferences are irrational. For instance, if you are exploitable, there is a series of binary decisions – accept this bet or reject it – such that your credences lead you to prefer accepting each bet to rejecting it, but also to prefer the series of rejections to the series of acceptances, because the former dominates the latter. Thus, if $a_i$ is the act of accepting the $i$th bet and $b_i$ is the act of rejecting it, your preferences are: $a_1 \succ b_1, \ldots, a_n \succ b_n$, but also $a_1 \ \& \ \ldots \ \& \ a_n \prec b_1 \ \& \ \ldots \ \& \ b_n$. And these, claim those who propose this response, are inconsistent in something close to the way that your beliefs are inconsistent if you believe each of $A_1, \ldots, A_n$, but also $\neg(A_1 \ \& \ \ldots \ \& \ A_n)$. What's more, they claim that inconsistent preferences are irrational in much the same way that inconsistent beliefs are irrational.

The problem with this response is that it isn't clear what makes inconsistent preferences irrational. As I see it, there are two possibilities, and neither is satisfactory.

First possibility: traditionally, in decision theory, what is thought to be irrational about inconsistent preferences is that they are exploitable (Davidson et al., 1955). That is, there are decision problems you might face where they'll lead you to choose poorly. For instance, cyclical preferences – such as $a_3 \prec a_1 \prec a_2 \prec a_3$ – are thought to be irrational because I can offer you $a_3$, then offer to let you switch to $a_1$ for a small price, then offer to let you

---

[26] See, for instance, Skyrms (1987), Howson and Urbach (1989), Christensen (1991), Armendt (1993), Christensen (1996), and Mahtani (2015). Vineberg (2001) offers a different objection to this proposal from mine.

switch to $a_2$ for a further small price, and finally offer to let you switch to $a_3$ for a yet further small price. Because of your preferences, you'll always pay the price for the switch, and that will leave you worse off than had you simply taken $a_3$ the first time it was offered and stuck with it for free. However, in this context, we cannot appeal to such an argument, for it clearly begs the question against the objection to which this is a response.

Second possibility: this account of the irrationality of inconsistent preferences draws on the case of beliefs. What is irrational about believing each proposition in an inconsistent set? Many epistemologists answer as follows: the aim of belief is truth, or at least a belief is better when it is true; and, if you believe each of an inconsistent set of propositions, you can know a priori that not all of your beliefs are true. However, as Kenny Easwaran and Branden Fitelson have pointed out, this is not a satisfactory argument (Easwaran & Fitelson, 2015; Easwaran, 2016). As the Miners Paradox shows, there is nothing irrational about an act that stands no hope of obtaining the highest possible utility (Parfit, ms; MacFarlane & Kolodny, 2010). In that paradox, we have three available acts and two states of the world. The payouts are as follows:

|               | Shaft A | Shaft B |
|---------------|---------|---------|
| Flood A       | $-10$   | $0$     |
| Flood B       | $0$     | $-10$   |
| Flood neither | $-1$    | $-1$    |

Now, for many people, the third option is at least rationally permissible. But, of course, it precludes the possibility of obtaining the best possible utility, namely, 0. The same goes for beliefs. *Flood A* and *Flood B* both give some hope of obtaining the highest possible utility, but they also open you up to the risk of the lowest; *Flood neither* does not. Similarly, *believing A*, *believing B*, and *disbelieving* $\neg(A \& B)$ gives some hope of getting things exactly right, but also some chance of getting things exactly wrong, while *believing A*, *believing B*, and also *believing* $\neg(A \& B)$ gives no hope of getting things exactly right, but also no chance of getting things exactly wrong – at least one of your beliefs is true in any world.

Now, it isn't obvious what it is to which preferences stand as beliefs stand to truth. Perhaps it is satisfaction. A preference $a \prec b$ is satisfied at a world if you receive $b$ rather than $a$ at that world. In that case, a set of preferences is inconsistent if there is no world at which all are satisfied. On that interpretation, the preferences to which non-probabilistic credences give rise via (DB1) in Dutch Book decision problems are inconsistent. It isn't possible to receive

$a_1$ instead of $b_1$, ..., $a_n$ instead of $a_n$, but also to receive $b_1$ & ... & $b_n$ instead of $a_1$ & ... & $a_n$. However, just as there is no reason to think that inconsistent beliefs are irrational just because they cannot all be correct, so there is no reason to think inconsistent preferences are irrational just because they cannot all be satisfied. The preferences your credences require when faced with the Dutch Book decision problems can never all be satisfied. But, unlike consistent preferences, they can never all be dissatisfied either. It isn't obvious, then, that having them is irrational.

## 6.3 A New Pragmatic Argument for Probabilism

Building on the insights of Schervish and Levinstein, we will now offer a new sort of pragmatic argument for the laws of credence that we have been considering. We focus primarily on Probabilism, but explain how to extend the argument to establish the Principal Principle and Conditionalisation as well.

In our argument for Probabilism, we build a utility function that measures how good a credence function is as a guide to action in a given possible world. Then we show that, if your credence function is not probabilistic, then there is an alternative that is probabilistic and that has greater utility at every possible world – that is, this alternative is guaranteed to act as a better guide to action than your current credence function. What's more, the same is not true if your credence function is probabilistic.

### 6.3.1 The Pragmatic Utility of an Individual Credence

Our first order of business is to create a utility function that measures how good an individual credence is as a guide to decision-making. Then we'll take the utility of a whole credence function to be the sum of the utilities of the individual credences that comprise it.

Suppose you assign credence $p$ to proposition $A$. Our job is to say how good this credence is as a guide to action. First, a little terminology:

- An $A$-act is an act that has the same utility at all worlds at which $A$ is true, and the same utility at all worlds at which $A$ is false.
- An $A$-decision problem is a decision problem in which all of the available acts are $A$-acts.
- Let $\mathcal{D}_A$ be the set of all $A$-decision problems.

We suppose that there is a probability function $P$ that says how likely it is that the agent will face different $A$-decision problems – since the set of $A$-decision problems is uncountably infinite, we actually take $P$ to be a probability density function, so that it does not give the probability of the agent facing a particular individual $A$-decision problems, but rather the probability of the agent facing some $A$-decision problem or other from a particular set. The idea here is that $P$ is something like an objective chance function.

With that in hand, we take the pragmatic utility of credence $p$ in proposition $A$ at a particular world $w$ to be the expected utility at $w$ of the choices that credence $p$ in $A$ will lead you to make when faced with the decision problems you encounter. For instance, if there were just two $A$-decision problems that you might encounter, then the pragmatic utility of $p$ at $w$ would be the sum of the utility at $w$ of the act that $p$ would require you to choose from the first decision problem, weighted by the probability you'd encounter that decision problem, and the utility at $w$ of the act that $p$ would require you to choose from the second decision problem, weighted by the probability you'd encounter that decision problem.

More generally, and a good deal more technically, the pragmatic utility of $p$ at $w$ is the integral, relative to the probability function $P$ over the possible $A$-decision problems $D$ in $\mathcal{D}_A$ you might face, of the utility at $w$ of the act you'd choose from $D$ using your credence $p$. Given $D$ in $\mathcal{D}_A$, let $D^p$ be the act you'd choose from $D$ using $p$ – that is, $D^p$ is one of the acts in $D$ that maximises expected utility by the lights of $p$. Thus, for any $D$ in $\mathcal{D}_A$, and any act $a$ in $D$,

$$\text{EU}_{p,u}(a) \leq \text{EU}_{p,u}(D^p)$$

where $\text{EU}_{p,u}(a) = pu(a \,\&\, A) + (1-p)u(a \,\&\, \overline{A})$, $a \,\&\, A$ is the outcome of $a$ when $A$ is true, $a \,\&\, \overline{A}$ is the outcome of $a$ when $\overline{A}$ is true, and so on. Then we define the pragmatic utility of having credence $p$ in proposition $A$ when $A$ is true as follows:

$$g_A(1,p) = \int_{\mathcal{D}_A} u(D^p \,\&\, A) dP$$

And we define the pragmatic utility of credence $p$ in $A$ when $A$ is false as follows:

$$g_A(0,p) = \int_{\mathcal{D}_A} u(D^p \,\&\, \overline{A}) dP$$

These are slight modifications of Schervish's and Levinstein's definitions.

### 6.3.2 g Is a Strictly Proper Scoring Rule

Our next order of business is to show that this utility function for $g$ has a particular important feature – in the jargon, it is a *strictly proper scoring rule* (Savage, 1971; Predd et al., 2009). Suppose I have a credence $p$ in $A$; and suppose you have credence $q$ in $A$. Then I can use my credence to assess how well I think your credence would serve me by taking the expected prag-matic utility of your credence from the point of view of my credence – that is, $EU_{p,g_A}(q) = pg_A(1,q) + (1-p)g_A(0,q)$. We say that $g$ is a *strictly proper scoring rule* if, for every credence $p$ in $A$, of all the possible credences in $A$, $p$ expects itself to do best. That is, if $p \neq q$, then $EU_{p,g_A}(q) < EU_{p,g_A}(p)$. We now show that, if we make a particular assumption discussed below, the pragmatic utility function $g_A$ from above is a strictly proper scoring rule. First, note:

$$EU_{p,g_A}(q) = pg_A(1,q) + (1-p)g_A(0,q)$$
$$= p\int_{\mathcal{D}_A} u(D^q \ \& \ A)dP + (1-p)\int_{\mathcal{D}_A} u(D^q \ \& \ \overline{A})dP$$
$$= \int_{\mathcal{D}_A} pu(D^q \ \& \ A) + (1-p)u(D^q \ \& \ \overline{A})dP$$
$$= \int_{\mathcal{D}_A} EU_{p,u}(D^q)dP$$

Now, by the definition of $D^q$, for all $D$ in $\mathcal{D}_A$,

$$EU_{p,u}(D^q) \leq EU_{p,u}(D^p)$$

So

$$EU_{p,g_A}(q) \leq EU_{p,g_A}(p)$$

What's more, if we make a particular assumption about the chance that you will face certain bets, then the inequality is guaranteed to be strict. For two credences $p$ and $q$ in $A$, we say that a set of decision problems *separates* $p$ and $q$ if (i) each decision problem in the set contains only two available acts, and (ii) for each decision problem in the set, $p$ expects one act to have higher expected value and $q$ expects the other to have higher expected value. Then, as long as there is some set of decision problems such that (i) that set separates $p$ and $q$ and (ii) $P$ assigns positive probability to this set, then

$$EU_{p,g_A}(q) < EU_{p,g_A}(p)$$

So if we assume this for all $p, q$, the scoring rule $g_A$ is strictly proper. This is a substantial assumption that places conditions on the utility function $u$ and the probability function $P$. While I have no space to consider how to justify it, it is important to note that it is the main source of concern about the current argument.

### 6.3.3 The Pragmatic Utility of a Whole Credence Function

The pragmatic utility function $g_A$ we have just defined assigns pragmatic util-ities to individual credences in $A$. In the next step, we define $G$, a pragmatic utility function that assigns pragmatic utilities to whole credence functions. We do this in the most obvious way – we take the utility of a credence function to be the sum of the utilities of the individual credences it assigns. Suppose $c : \mathcal{F} \to [0, 1]$ is a credence function defined on the set of propositions $\mathcal{F}$. Then:

$$G(c, w) = \sum_{X \in \mathcal{F}} g_A(w(A), c(A))$$

where $w(A) = 1$ if $A$ is true at $w$ and $w(A) = 0$ if $A$ is false at $w$. In this situation, we say that $G$ is generated from the scoring rules $g_A$ for each $A$ in $\mathcal{F}$.

### 6.3.4 Predd et al.'s Dominance Result

Finally, we appeal to a theorem due to Predd et al. (2009):

**Theorem 6.1** (Predd et al. 2009)   *Suppose G is generated from strictly proper scoring rules $g_A$ for A in $\mathcal{F}$. Then:*

(I) *If c is not a probability function, then there is a probability function $c^\star$ such that $G(c, w) < G(c^\star, w)$ for all worlds w.*
(II) *If c is a probability function, then there is no credence function $c^\star \neq c$ such that $G(c, w) \leq G(c^\star, w)$ for all worlds w.*

This furnishes us with a new pragmatic argument for Full Probabilism. In this argument, we see that non-probabilistic credences are irrational not because there is some series of decision problems such that, when faced with them, the credences require you to make a dominated series of choices. Rather, they are irrational because there are alternative credences that dominate them – however the world turns out, the expected or average utility you'll gain from making decisions using those alternative credences is greater than the expected or average utility you'll gain from making decisions using the original credences.

We can also give such pragmatic utility arguments for the Full Principal Principle and Plan Conditionalisation as well, though not for (Full) Regularity. By analogy with Theorem 6.1, Pettigrew (2013) shows that, if a credence function violates the Full Principal Principle, there is an alternative that has greater expected pragmatic utility relative to any epistemically possible objective chance function, provided the pragmatic utility function is generated by strictly proper scoring rules. And, under a similar assumption, Briggs and Pettigrew (2018) show that, if you have a prior credence function and an updating rule that violate Plan Conditionalisation, then there is an alternative prior and updating rule that has greater combined pragmatic utility than your prior and updating rule at every epistemically possible world.

## 7 Generalising the Dutch Book Arguments

So far in this Element, our arguments have been based on at least three assumptions about the framework in which they take place. First, we have assumed that the propositions to which you assign credences are *uncentered propositions*. That is, they might have different truth values at different possible worlds, but they always have the same truth value at different times, different places, and relative to different individuals within a given world. Second, we have assumed that these propositions are governed by classical logic. Thus, every proposition is true or false; a disjunction is true exactly when one disjunct or the other or both are true, and false otherwise; and so on. And, third, we have assumed that your beliefs are accurately represented by a single credence function that assigns to each proposition you consider a precise numerical value. We conclude, in this final section, by asking what happens when we change these assumptions.

### 7.1 Self-Locating Credences

First, let us allow that the propositions to which we assign credences might be *centred* or *self-locating*. In particular, we will explore what happens when we allow that you might have a credence in a proposition whose truth value in some possible worlds changes from one time to another. Let's begin with an example.

Beatty has volunteered to participate in a sleep study. On Sunday night, he will be put to sleep. Then a fair coin will be tossed. If the coin lands heads, the experimenter will wake Beatty on Monday, then put him back to sleep until the experiment is over. If, on the other hand, the coin lands tails, she'll wake him on Monday, put him back to sleep, erase his memories of his Monday awakening, wake him up again on Tuesday in an identical setting, then put him back to sleep until the experiment is over. Beatty knows all of this – he was told it all before he signed the consent form. On Sunday, Beatty's credence that the fair coin will land heads should be $\frac{1}{2}$ (by the Principal Principle). What should it be when he is woken on Monday?[27]

Our first response might be to look to the conclusions of Section 4 for advice. After all, is this not a question about how we should respond to new evidence? And don't we know from that section that we should plan to respond to new evidence by conditionalising on it? Well, yes and no. Our question does indeed concern how we should respond to evidence, but Plan Conditionalisation won't

---

[27] This is a version of the Sleeping Beauty puzzle (Elga, 2000). It is closely related to the Absent-Minded Driver's Paradox (Piccione & Rubinstein, 1997) from the economics literature.

help us here. Suppose $H_1$ is the centred proposition that says that it is Monday and the coin landed Heads; $T_1$ says it is Monday and it landed Tails; and $T_2$ says it is Tuesday and it landed Tails. When Beatty wakes up on Monday, his total evidence is the centred proposition $H_1 \vee T_1 \vee T_2$. But his credence in that proposition on Sunday was 0, since he was then certain that it was Sunday and not Monday or Tuesday. So conditionalisation imposes no constraints on his credences on Monday. But clearly there are such rational constraints – it would be irrational, for instance, for Beatty to become certain in Tails when he wakes up on Monday. So what are these constraints and can we use Dutch Book arguments to establish them?

Those who consider this problem tend to divide into two camps. Halfers hold that Beatty should retain his credence $\frac{1}{2}$ in Heads when he awakens on Monday. They might argue, for instance, that Beatty knows on Sunday exactly what evidence he'll obtain when he wakes on Monday. Thus, when he does wake on Monday, he learns nothing new. So he should continue to match his credence in Heads to the known objective chance of Heads (Lewis, 2001). Thirders, in contrast, hold that Beatty's credence in Heads should be $\frac{1}{3}$. They might argue, for instance, that (i) when he awakens on Monday he should have the same credences in $T_1$ and $T_2$, by a weakened version of the Principle of Indifference, which says that if your evidence doesn't tell between two times within the same uncentred world, you should have equal credence in both, and (ii) on Monday his conditional credence in Heads ($H_1$) given that it is Monday ($H_1 \vee T_1$) should be $\frac{1}{2}$ (Elga, 2000). This ensures that his Monday credence in $H_1$ should be $\frac{1}{3}$.

Can we use sure loss arguments to resolve the disagreement between Thirders and Halfers? Here's an initial attempt. As we noted above, the evidence Beatty receives upon awakening on Monday is exactly what he knew he would receive. Now, in the uncentred case, we can devise a straightforward Dutch Strategy against anyone who changes their credence in any way when they have learned nothing that they did not know they would learn. Let's apply this here: If Beatty's credence in Heads decreases between Sunday and Monday, then he'll buy a £1 bet on Heads on Sunday for more than he'll sell that same bet for on Monday, losing him money overall at all epistemically possible worlds; if instead his credence increases, then he'll sell a £1 bet on Heads on Sunday for less than he'll buy it for on Monday, again losing him money overall. For instance, suppose Beatty is a Thirder, so that his credence in Heads on Monday is $\frac{1}{3}$. Then he should buy a £24 bet on Heads for £10 on Sunday, and he should sell a £24 bet on Heads for £9 on Monday. And this is guaranteed to lose him £1.

This seems to tell in favour of the Halfer solution to our puzzle. But there's a problem. Deterministic Dutch Strategies are used to show that certain priors together with certain deterministic updating rules are irrational. A deterministic

|  | *Sun* | *Mon* | *Tues* | Net gain |
|---|---|---|---|---|
| *Heads* | $-10 + 24$ | $9 - 24$ | $\times$ | $-1$ |
| *Tails* | $-10$ | $9$ | $\times$ | $-1$ |

updating rule specifies the different possible evidence that you might acquire, and then for each possible piece of evidence, it specifies the credence function you will adopt if you obtain that piece of evidence. A Dutch Strategy against a prior and an updating rule specifies some initial bets that your prior credences require you to take, and then it assigns to each possible piece of evidence you might acquire some bets that your credences will require you to accept if you obtain that evidence and update in the way the rule recommends. Now, the problem is that the sure loss bets described above aren't a Dutch Strategy in this sense. It is crucial to the success of the betting strategy described above that Beatty is only offered the bets in question *once* if Heads and *once* if Tails, even though he receives exactly the same evidence *once* if Heads, but *twice* if Tails – if Tails, he receives $H_1 \vee T_1 \vee T_2$ as his evidence when he wakes up on Monday *and* when he wakes up on Tuesday. If he were offered the bets twice if Tails – once on Monday and once on Tuesday – then assuming that his Monday and Tuesday credences are the same, he would lose £1 in the Heads world, as before, but he would gain £8 in the Tails world, for he would pay out £10 for the bet on Sunday and receive nothing in return, take in £9 for the bet on Monday and give out nothing in return, and take in £9 for the bet on Tuesday and give out nothing in return, thus leaving him £8 richer. This shows that the betting strategy against Thirder Beatty is not a Dutch Strategy. It dictates that Beatty is offered different bets in the Tails world on Monday (selling the £24 bet for £9) and on Tuesday (no bets), even though his evidence is identical on those two days. Thus, it is defined on a more fine-grained set of possibilities than the set on which Beatty's updating rule is defined. And so the fact that it leads to a sure loss for Beatty does not impugn his rationality. A true Dutch Strategy against Beatty would have to offer him exactly the same bets in each possible centred world in which Beatty receives the same evidence – that is, on Monday in the Heads world, on Monday in the Tails world, and on Tuesday in the Tails world.

With this understanding of Dutch Strategies in centred cases, we now notice that it is in fact Halfer Beatty rather than Thirder Beatty who is vulnerable to a Dutch Strategy. On Sunday, he'll sell a £24 bet on Heads for £13. And then whenever he awakens, he'll buy a £18 bet on Heads for £8. Thus, he'll lose £3 for sure:

|        | *Sun*     | *Mon*     | *Tues*   | Net gain |
|--------|-----------|-----------|----------|----------|
| *Heads* | $13 - 24$ | $-10 + 18$ | $\times$ | $-3$     |
| *Tails* | $13$      | $-8$      | $-8$     | $-3$     |

Now, note that one consequence of this is that how you should update is not a function only of the evidence that you receive, but also of the number of centres within each uncentred possible world at which you receive that evidence.

But we aren't finished yet! As Frank Arntzenius (2002) noted and R. A. Briggs (2010) developed in detail, whether or not the Dutch Strategies work against a non-Thirder in Beatty's position depends on the decision theory we use to evaluate the bets. The two main decision theories are evidential decision theory (abbreviated EDT and introduced by Richard Jeffrey (1983)) and causal decision theory (abbreviated CDT and introduced by David Lewis (1981) and others, including Jim Joyce (1999)). According to both, when we face a decision problem, we should choose an option that maximises expected utility. But they disagree on the probabilities that we should use to calculate the expected utility of a particular option. According to the CDTist, the relevant probability for a given state of the world measures how likely choosing that option makes that state of the world, while for the EDTist, it is how likely you would take that state to be were you to learn that you chose that option. Suppose, for instance, that there is a gene that leads you to choose chocolate ice cream over strawberry whenever the choice presents itself, but the gene also causes you to develop a dreadful medical condition. You're faced with a choice between chocolate and strawberry ice cream. EDT tells you to choose the strawberry ice cream, since if you were to learn that you chose the chocolate, that would increase your credence that you have the dreadful medical condition, which has very low utility, and thus choosing the chocolate has low expected utility for the EDTist. CDT, on the other hand, tells you to choose whichever of the two you prefer, since whichever you choose isn't going to have any causal effect on whether or not you have the gene, and so choosing one or the other isn't going to make it more or less likely that you have it.

Now, consider the Dutch Strategy against Halfer Beatty. We assumed above that, if he has credence $\frac{1}{2}$ in Heads on Monday, he should pay £8 for an £18 bet on Heads. This follows from (DB1), which we've been assuming for much of this Element. But the EDTist might deny this. After all, suppose Beatty were to learn that he accepts the bet. Then we might reasonably suppose that he also thereby learns that he will accept it at any time at which he has the same credences. So the utility of accepting this bet in the Heads world is the utility of accepting it once there, which is the utility of gaining £10. But the utility of accepting it in the Tails world is the utility of accepting it twice, which is the utility of losing £16. And, assuming that learning that he accepts the bet doesn't affect his credences in Heads and Tails, the expected utility of accepting the bet is therefore $\left(\frac{1}{2} \times 10\right) + \left(\frac{1}{2} \times -16\right) = -3 < 0$. So, Beatty should not accept it. And it turns out that, if he is an EDTist, only Halfer Beatty is not vulnerable to a Dutch Strategy.

As Briggs shows, the upshot of all of this is:

- EDTists should be Halfers in Beatty's situation – if they are not, there is a Dutch Strategy against them, while if they are, there is not.
- CDTists should be Thirders in Beatty's situation – if they are not, there is a Dutch Strategy against them, while if they are, there is not.

So Dutch Strategy arguments do not resolve the Sleeping Beauty puzzle, which Beatty's case dramatises – but they do show how the two rival putative solutions are related to two rival solutions to the general problem of how to choose rationally. The following theorem generalises a result due to Briggs (2010):

**Theorem 7.1** *Suppose:*

- $\mathcal{W}$ *is the set of possible uncentred worlds;*
- *c is your credence function and E is your evidence at one time;*
- $E_1, \ldots, E_n$ *are the pieces of evidence you might receive at some future time;*
- $E_1, \ldots, E_n$ *are disjoint;*
- *for each uncentred world w, there is a unique $E_i$ such that $E_i$ is true at some time in w;*
- *if you receive $E_i$, your credence function at that future time will be $c_i$;*
- *N(w) is the number of centres in uncentred world w at which you receive evidence E;*
- *$N_i(w)$ is the number of centres in uncentred world w at which you receive evidence $E_i$.*

*Then:*

(I) *If you are a CDTist, then it ought to be that, for all $1 \leq i \leq n$ and all w in $\mathcal{W}$,*

$$\frac{\frac{c_i(w)}{N_i(w)}}{\sum_{w' \cap E_i \neq \emptyset} \frac{c_i(w')}{N_i(w')}} = \frac{\frac{c(w)}{N(w)}}{\sum_{w' \cap E \neq \emptyset} \frac{c(w')}{N(w')}}$$

*If not, there is a Dutch Strategy against you; if so, there isn't.*

(II) *If you are an EDTist, then it ought to be that, for all $1 \leq i \leq n$ and all $w$ in $\mathcal{W}$,*

$$c_i(w) = \frac{c(w)}{\sum_{w' \cap E_i \neq \varnothing} c(w')}$$

*If not, there is a Dutch Strategy against you; if so, there isn't.*

We sketch the proof in Section 8.

## 7.2 Non-classical Credences

As we have stated it, the Dutch Book arguments assume at a number of points that the logic that governs the propositions to which we assign our credences is classical logic. In this section, we ask what happens if we change this assumption.

For instance, at the heart of the Dutch Book argument is the notion of a bet on a proposition. Now, as we have defined it, a £$S$ bet on a proposition $A$ pays out £$S$ if $A$ is true and £0 if $A$ is false. But we have said nothing of what the payout will be if $A$ takes some alternative truth value. Suppose Admiral Howard is considering the proposition $B$ that there will be a battle at sea tomorrow. $B$ is a future contingent, and on some accounts it will take truth value *neither* (N), and it will be governed by strong Kleene logic (which we'll describe in greater detail below). How much does a £$S$ bet on $B$ pay out if it takes truth value N? Or suppose Leah is considering the Liar sentence $L$. On some accounts, $L$ takes the truth value *both* (B), and it is governed by the Logic of Paradox (which again we'll describe below). How much does a £$S$ bet on $L$ pay out if it takes truth value B? In Section 7.2.1, we'll consider how to reformulate the Dutch Book argument to accommodate such logics.

As well as assuming classical logic when we defined the bets that sit at the heart of the Dutch Book arguments, we also assumed it when we came to prove that certain combinations of bets are guaranteed to yield a loss. Consider Safet, who is entertaining the proposition $R$, which says that the handkerchief in his pocket is red. And suppose that *red* is a vague concept. Suppose he assigns the following credences:

| $R$ | $\overline{R}$ | $R \vee \overline{R}$ | $R \& \overline{R}$ |
|-----|-----|-----|-----|
| 0.2 | 0.2 | 1 | 0 |

Then his credence in the disjunction $R \vee \overline{R}$ is greater than his credence in the exclusive disjuncts $R$ and $\overline{R}$. In our Dutch Book argument, in order to show that Safet is irrational, we assumed that his credences should lead him to buy a £100 bet on $R \vee \overline{R}$ for £90, sell a £100 bet on $R$ for £30, and sell a £100 bet on $\overline{R}$ for £30. And we argued that, taken together, these would be guaranteed to lose him money. After all, we said, the bet on $R \vee \overline{R}$ pays out iff exactly one of the bets on $R$ and $\overline{R}$ pay out. So, in all epistemically possible worlds, Safet's gains and losses from the payouts of the bets will cancel each other out, and he'll end up gaining £60 in total from the two bets he sold, but losing £90 from the bet he bought – thus, he'll lose £30 overall. But this argument assumes that $R \vee \overline{R}$ is true iff exactly one of $R$ and $\overline{R}$ is true. The propositions $R$ and $\overline{R}$ involve vague concepts. On some accounts, these are governed by a supervaluationist semantics. And, on such a semantics, this assumption does not hold. On such a semantics, while the concept *red* is vague – so that some items neither determinately fall under it nor determinately fall outside it – there is a set of admissible precisifications of it that are not vague – on each precisification, an item either falls under the concept or it does not. And we say that the proposition is true if it is true on all precisifications. Now, since Safet's handkerchief is a borderline case, it will fall under the concept *red* on some precisifications and not on others. But, on all precisifications, it will fall under the concept or it won't. Thus, neither $R$ nor $\overline{R}$ will be true on this account, but $R \vee \overline{R}$ will be. So Safet will pay out £90 to buy the bet on $R \vee \overline{R}$, receive £30 for selling the bet on $R$ and £30 for the bet on $\overline{R}$; he will receive £100 from the bet he bought and will have to pay out nothing from either of the bets he sold. So he will end up with a net gain of £70. So the classical assumption of the Dutch Book argument fails. In Section 7.2.2, we'll consider how to reformulate the argument to accommodate such a semantics.

### 7.2.1 Many-Valued Logics

Let's begin with strong Kleene logic – the logic that is sometimes thought to govern future contingents, such as Admiral Howard's $B$ – and Logic of Paradox – the logic that is sometimes thought to govern semantic paradoxes, such as Leah's $L$. According to classical logic, there are just two truth values, *true* (T) and *false* (F). And the truth values of negations, conjunctions, and disjunctions are determined as the following truth tables demand:

| $A$ | $\overline{A}$ | | $\&$ | T | F | | $\vee$ | T | F |
|-----|-----|---|------|---|---|---|--------|---|---|
| T | F | | T | T | F | | T | T | T |
| F | T | | F | F | F | | F | T | F |

In classical logic, one proposition $A$ is a logical consequence of another $B$ (written $X \models_{cl} Y$) iff whenever $A$ takes truth value T, $B$ takes truth value T as well. And a proposition $A$ is a tautology (written $\models_{cl} X$) – that is, $A$ takes T in all worlds.

According to strong Kleene logic and Logic of Paradox, in contrast, there are three truth values, *true* (T), *false* (F), and a third truth value. In strong Kleene logic, this is called *neither* (N) and it is interpreted as saying that the proposition is neither true nor false; in Logic of Paradox, this is called *both* (B) and it is interpreted as saying that the proposition is both true and false. The truth tables are the same for both logics. I'll give them in terms of T, F, and N here. The Logic of Paradox truth tables are obtained by replacing N with B throughout.

| $A$ | $\overline{A}$ |   | $\&$ | T | N | F |   | $\vee$ | T | N | F |
|-----|-----|---|------|---|---|---|---|--------|---|---|---|
| T | F |   | T | T | N | F |   | T | T | T | T |
| N | N |   | N | N | N | F |   | N | T | N | N |
| F | T |   | F | F | F | F |   | F | T | N | F |

In strong Kleene logic, one proposition $B$ is a logical consequence of another $A$ (written $X \models_{sk} Y$) iff whenever $A$ takes T, $B$ takes T as well. So a proposition $A$ is a tautology (written $\models_{sk} A$) iff $A$ takes T in all worlds, and $A$ is a contradiction (written $A \models_{sk}$) iff, in each world, $A$ does not take T. In Logic of Paradox, on the other hand, $B$ is a logic consequence of $A$ (written $A \models_{lp} B$) iff whenever $A$ takes one of T or B, $B$ takes one of T or B. So a proposition $A$ is a tautology (written $\models_{lp} A$) iff, in each world, $A$ takes either T or B, while $A$ is a contradiction (written $A \models_{sk}$) iff, in each world, $A$ does not take T or B. Thus, the difference between strong Kleene logic and Logic of Paradox lies in what we call the *designated values*. In a many-valued logic $\mathcal{L}$, the set of designated truth-values $D_{\mathcal{L}}$ is the subset of the set of all truth-values for which the following holds: $B$ is a logic consequence of $A$ iff for each world, if the truth value of $A$ is designated at that world, then the truth value of $B$ is designated as well. Thus, $D_{cl} = \{T\}$, $D_{sk} = \{T\}$, and $D_{lp} = \{T, B\}$.

In order to formulate Dutch Book arguments when the background logic is strong Kleene or Logic of Paradox or some other many-valued logic, we must specify the payout of a £$S$ bet on $A$, when $A$ takes truth-value N or B or some other truth value. Now, this is a matter of definition, so it is in our power to determine it in any way we please. But we must ensure that, however we determine it, we know what price your credence in $A$ rationally requires you to pay for a £$S$ bet on $A$. One natural suggestion is this:[28]

> payout of £$S$ bet on $A$ at world $w$ =
> $$\begin{cases} £S & \text{if } A \text{ takes a designated truth value at } w \\ £0 & \text{otherwise} \end{cases}$$

And if your credence in $A$ is $p$, you are rationally required to pay £$x$ for a £$S$ bet on $A$ iff $x < pS$, and rationally permitted to pay £$x$ iff $x \leq pS$. On this understanding, your credence in $A$ is really a credence that $A$ takes a designated truth value. Thus, in classical logic and strong Kleene logic, it is your credence that $A$ takes T, whereas in Logic of Paradox, it is your credence that $A$ takes T or B. In all three cases, we might say that it is your credence that $A$ is true, where that covers both T and B, but not N or F.

With this in hand, we can ask: which sets of credences over propositions governed by a many-valued logic are finitely exploitable? For a wide range of many-valued logics, though not all, an elegant result due to Jeff Paris (2001) gives the answer.[29]

First, a definition. Recall: we say that a set of propositions is an *algebra* if it is closed under negation, conjunction, and disjunction. We also say that a set of propositions is a *lattice* if it is closed under conjunction and disjunction. And, given a set of propositions $\mathcal{F}$, we write $\mathcal{F}^\dagger$ for the smallest lattice that contains $\mathcal{F}$.

Second, a law of credence.

> **Full Generalised Probabilism** Suppose $\mathcal{F}$ is the set of propositions to which you assign credences, and $E$ is the strongest proposition you have as evidence. Suppose many-valued logic $\mathcal{L}$ governs the propositions in $\mathcal{F}$. Then: it should be possible to extend your credences over $\mathcal{F}$ to credences over $(\mathcal{F} \cup \{E\})^\dagger$, so that the extended credences satisfy the following conditions:
>
> (FGP1a) If $E \models_\mathcal{L} X$, then $c(X) = 1$;
> (FGP1b) If $E, X \models_\mathcal{L}$, then $c(X) = 0$;
> (FGP2) If $E, X \models_\mathcal{L} Y$, then $c(X) \leq c(Y)$;
> (FGP3) $c(X \vee Y) = c(X) + c(Y) - c(X \& Y)$.

Notice that, if $\mathcal{L}$ is classical logic, then Full Generalised Probabilism is equivalent to Full Probabilism. From Finite Additivity and Normalisation, we can derive (FGP1-4), and vice versa.

Third, Paris' theorem.

---

[28] For alternatives, see Williams (2012a).
[29] See also Williams (2012a,b).

**Theorem 7.2**   *Suppose $\mathcal{F}$ is the set of propositions to which you assign credences. Suppose many-valued logic $\mathcal{L}$ governs the propositions in $\mathcal{F}$. And suppose the following holds of the truth values in $\mathcal{L}$: at each world,*

> (T2) *$A$ & $B$ takes a dedicated truth value iff $A$ and $B$ both take dedicated truth values;*
> (T3) *$A \lor B$ takes a non-dedicated truth value iff $A$ and $B$ both take non-dedicated truth values.*

*Then, your credences violate Full Generalised Probabilism iff they are finitely strongly fully exploitable.*

This theorem, together with the relevant versions of (DB1) and (DB3), gives the Dutch Book argument for Full Generalised Probabilism. We sketch the proof in Section 8.[30]

### 7.2.2 Supervaluationist Semantics

Recall Safet from above. $R$ says that his handkerchief is red; and *red* is a vague concept. On a supervaluationist semantics for $R$, there is a set of legitimate precisifications of *red*. $R$ is true at a world if it is true at that world on all precisifications; false if false on all precisifications; and neither true nor false otherwise. In general, given a set of propositions $\mathcal{F}$, each legitimate precisification takes each world and returns a complete classical assignment of truth values to the propositions in $\mathcal{F}$. A proposition is true at a world iff it is true at that world relative to every precisification.

Now, as in the previous section, we must specify (i) what the payout of a £$S$ bet on $A$ is at a possible world; and (ii) what price your credence in $A$ rationally requires you to pay for a £$S$ bet on $A$. The natural suggestion:

payout of £$S$ bet on $A$ at world $w =$

$$\begin{cases} £\,S & \text{if } A \text{ is true on all legitimate precisifications at } w \\ £\,0 & \text{otherwise} \end{cases}$$

And, if your credence in $A$ is $p$, you are rationally required to pay £$x$ for a £$S$ bet on $A$ iff $x < pS$, and rationally permitted to pay £$x$ iff $x \leq pS$. On this understanding, your credence in $A$ is a credence that $A$ is true on all legitimate precisifications.

---

[30] Strong Kleene logic and the Logic of Paradox are two of the simplest non-classical logics. But there are Dutch Book arguments available for more sophisticated logics. See, for instance, Bradley (2017) and Gerla (2000).

As above, we can ask: which credences on propositions governed by such a supervaluationist semantics are finitely exploitable? As Jeff Paris (2001) noted, the answer is given by a neat observation due to Jean-Yves Jaffray (1989).[31]

First, a law of credence. Given a possible world $w$, we write $A_w$ for the strongest proposition that is true at $w$ on all precisifications. We call such propositions state descriptions, since a proposition is true at $w$ iff it is classically entailed by $A_w$.

> **Supervaluational Dempster-Shaferism** Suppose $\mathcal{F}$ is the set of propositions to which you assign credences, and $E$ is the strongest proposition you have as evidence. Suppose a supervaluationist semantics governs the propositions in $\mathcal{F}$. Then: it should be possible to extend your credences over $\mathcal{F}$ to credences over $(\mathcal{F} \cup \{E\})^*$, so that the extended credences satisfy the following conditions:[32]
>
> (DS1a)  If $E \models_{cl} A$, then $c(A) = 1$;
> (DS1b)  If $E, A \models_{cl}$, then $c(A) = 0$;
> (DS2)   If $E, A \models_{cl} B$, then $c(A) \le c(B)$;
> (DS3)   For any proposition $A$ in $\mathcal{F}$,
>
> $$c(A) \ge \sum_{B \subsetneq A} (-1)^{|A-B|+1} c(B)$$
>
> (DS4)  For any proposition $A$ in $\mathcal{F}$ that is not a state description for any world,
>
> $$c(A) = \sum_{B \subsetneq A} (-1)^{|A-B|+1} c(B)$$
>
> A credence function that satisfies (DS1-3) is known as a *Dempster-Shafer belief function*. If it satisfies (DS4) as well, we say that it is a *supervaluational Dempster-Shafer belief function* relative to the set of legitimate precisifications.

Thus, Supervaluational Dempster-Shaferism requires your credence function to be a particular sort of Dempster-Shafer belief function (Dempster, 1968; Shafer, 1976). Belief functions were originally introduced by Dempster and developed by Shafer to model credences as measures of the evidence that an individual has for a proposition. Here, we take no stance on what credences measure. We simply show that, if they are defined over propositions for which there is a supervaluationist semantics, they should satisfy Supervaluational Dempster-Shaferism to avoid being exploitable.

---

[31] Seamus Bradley (2017) generalises this answer to cases in which the legitimate precisifications are not classical assignments of truth values and the set of propositions is not a Boolean algebra.

[32] Here, since $(\mathcal{F} \cup \{E\})^*$ is a Boolean algebra, we represent the propositions in it as subsets of a set.

Second, a theorem.

**Theorem 7.3**  *Suppose $\mathcal{F}$ is the algebra of propositions to which you assign credences, and $E$ is the strongest proposition you have as evidence. Suppose a supervaluationist semantics governs the propositions in $\mathcal{F}$. Then, your credences violate Supervaluational Dempster-Shaferism iff they are finitely strongly fully exploitable.*

We sketch the proof in Section 8. This theorem, together with the relevant versions of (DB1) and (DB3), gives a Dutch Book argument for Supervaluational Dempster-Shaferism.

## 7.3 Other Representations of Belief

Throughout this Element, we have assumed that you and the cast of characters from our examples have precise numerical credences. That is, we've assumed that your beliefs can be represented by a single mathematical function that takes each proposition about which you have an opinion, and returns a single numerical value that is intended to measure how strongly you believe that proposition. We have made that assumption largely because it is in the framework of such precise numerical credences that the Dutch Book argument has most often been discussed in the philosophical literature in the past. But similar arguments can be formulated when we represent beliefs in different ways. In this final section, we consider one related way, namely, imprecise credences.[33] We present the alternative representation, we describe a (partial) decision theory for individuals represented in this way, and we investigate which laws we can establish for such individuals using pragmatic arguments.

Imperia entertains the proposition *Medium*, which says that global mean surface temperature on Earth will rise by between 0°C and 1°C in the coming century. She also entertains the proposition *Gold*, which says that the price of gold will double in the coming decade. If we represent her by a credence function, she will assign a numerical credence to both of those propositions. And that will ensure that she is represented as either more or less or exactly as confident in *Medium* as in *Gold*. But it turns out that she is none of these things. Her levels of confidence in the two propositions are incomparable. She is not more confident in one than the other, and she is not exactly as confident in both. However, she does have some opinions about them. She judges each to

---

[33] There are, of course, other representations of beliefs, such as full beliefs, comparative confidences, or upper and lower probabilities. And there are Dutch Book arguments concerning each of them (Rothschild, 2019; Fishburn, 1986; Walley, 1991). Unfortunately, I have space to treat only one representation here.

be more than 20 per cent likely, and she judges them to be independent of one another. How are we to represent Imperia's beliefs faithfully? Instead of representing her by a single credence function $c$, we represent her using a set of credence functions **P**. Following van Fraassen (1990), we call **P** her *representor*.[34] It contains all and only the credence functions that make the judgments that she makes. That is, **P** contains $c$ iff

  (i) $c(Medium) > 0.2$,
 (ii) $c(Gold) > 0.2$, and
(iii) $c(Medium \mid Gold) = c(Medium)$.

There are a number of different putative laws that we might take to govern representors. And, for some of them, there are Dutch Book arguments. In this section, we'll consider three:

> **Consistency** Your representor **P** should contain at least one credence function.
> **Precision** Your representor **P** should contain at most one credence function.
> **Pointwise Probabilism** Every credence function in your representor **P** should obey Full Probabilism.

Nearly all proponents of imprecise credences endorse Consistency and Pointwise Probabilism. None endorse Precision. After all, it says that, while there might be individuals who are representable only by genuinely imprecise credences, they are irrational – the rational ones are represented by a single precise credence function. I include it here because there appears to be a Dutch Book argument in its favour.

To give a Dutch Book argument for any law governing representors, we need a decision theory for individuals represented in this way. But, while many decision theories have been proposed, all suffer from various problems. I'll present three popular varieties and explain which of the laws listed above we can justify by Dutch Book arguments using these decision theories.

> **Liberal** (or $E$-**admissibility**) (Levi, 1974; Seidenfeld et al., 1995; Williams, 2014; Moss, 2015) Faced with a decision problem, an available act $a$ is rationally permissible for an individual with representor **P** iff there is a credence function $c$ in **P** such that $a$ is rationally permissible for an individual with precise credence function $c$.

---

[34] For more on this framework, see Walley (1991); Levi (1974); Joyce (2010); Bradley (2016), and Moss (2018).

Now, based on Liberal, we can formulate Dutch Book arguments for all of our laws from above. They come in different varieties. First, some terminology: Just as we say that a credence function is *weakly fully exploitable* if there are bets, each of which it permits you to accept, that, taken together, provide a sure loss, so we say the same of representors. What's more, we also say that a representor is *weakly fully thwartable* if there are bets, each of which it does not permit you to accept, that, taken together, provide a sure gain.

First, the argument for Consistency. Since an option is permissible by the lights of a representor only if it contains a credence function that permits it, an empty representor $\mathbf{P}_\varnothing$ permits no options. In particular, it does not permit even a £1 bet on a necessary proposition $\top$ for free. That is, there is a single sure gain bet that it does not permit you to accept. Thus, $\mathbf{P}_\varnothing$ is *singly weakly fully thwartable*.

Next, the argument for Pointwise Probabilism. This piggybacks on the corresponding argument in the framework of precise credences. Suppose $\mathbf{P}$ contains a credence function $c$ that isn't a probability function. We must then choose our decision theory for non-probabilistic precise credence functions: (DB1) or (MSEU)? If (MSEU), then $c$ is singly strongly fully exploitable, so $\mathbf{P}$ is *singly weakly fully exploitable*. On the other hand, if (DB1), then $c$ is finitely strongly fully exploitable, so $\mathbf{P}$ is *finitely weakly fully exploitable*.

Finally, the Dutch Book argument for Precision (Elga, 2010). Suppose $\mathbf{P}$ contains at least two credence functions, $c \neq c'$, with $c(A) = r < r' = c'(A)$. Then find $x, y$ such that $r < x < y < r'$. Then $c(A)$ requires you to sell a £1 bet on $A$ for £$x$, while $c'(A)$ requires you to buy a £1 bet on $A$ for £$y$. The total net gain is £$(x - y)$ at all worlds, which is negative. Thus, $\mathbf{P}$ is *finitely weakly fully exploitable*.

Proponents of the imprecise credence framework respond to this latter argument in one of two ways:

(i)  They argue that it does not show that a representor that violates Precision is irrational; or
(ii) They accept that it does, but they lay the blame with the decision theory, not with the imprecise credences, and they introduce a different decision theory for representors.

Following (i): Sarah Moss (2015) argues that it is not always irrational to have beliefs that *permit* you to accept each of a series of bets that, taken together, lose you money for sure – that is, it is not always irrational to be finitely *weakly* fully exploitable. Seamus Bradley and Katie Steele argue for the same conclusion by a different route (Bradley & Steele, 2014).

Another possibility that falls under (i): we might say that, while this argument shows that you would be irrational if you were to make your decisions about both of the bets using the same representor, this will never be the case. On this proposal, once you make your decision whether to accept or reject the first bet, this changes your representor, so that you face the decision whether to accept or reject the second bet with a different one. One version of this proposal says that you face the decision whether or not to accept the first bet with the representor **P**; but, if you decide to accept that bet, then your representor changes so that it includes only those members of **P** for which accepting the first bet maximises expected utility; then you face the decision whether or not to accept the second bet with this new updated representor; and it will not permit you to choose to accept it (Williams, 2014). Against this, it is objected that this decision rule requires you to change your beliefs – as represented by your representor – even though you don't receive any new evidence that you take to pertain to them – you change your beliefs solely because you chose one way in a particular decision.

Following (ii): One possible alternative to Liberal is the *Maximin* (or Γ-*Maximin*) decision rule. It avoids the Dutch Book argument for Precision described above.

> **Maximin** (Gilboa & Schmeidler, 1989; Berger, 1985; Seidenfeld, 2004)
> Faced with a decision problem, an available act $a$ is rationally permissible for an individual with representor **P** iff for any alternative available act $a'$, the lowest expected utility for $a'$ among credence functions in **P** is at most the lowest expected utility for $a$ among credence functions in **P**.

Now, notice that, in the Dutch Book argument for Precision above, since $c'$ requires you to accept the second bet, it must require you to reject the first, in which case the expected utility of accepting the first relative to $c'$ is negative; and mutatis mutandis for $c$ and the second bet. So the worst-case expected utility for rejecting either of the bets is 0, which is greater than the worst-case expected utility for accepting either of them. Thus, Maximin will require you to reject both.

However, while there is no sure loss argument for Precision when we use Maximin, there is a sure gain argument (Elga, 2010). As before, suppose **P** contains at least two credence functions, $c \neq c'$, with $c(A) = r < r' = c'(A)$. Then, as before, find $x, y$ such that $r < x < y < r'$. Then $c(A)$ requires you not to buy a £1 bet on $A$ for £$x$, while $c'(A)$ requires you not to sell a £1 bet on $A$ for £$y$. So Maximin will require you to refuse both bets. But the total net gain of them is £$(y - x)$ at all worlds, which is positive. Thus, **P** is *finitely weakly fully thwartable*.

Also following (ii): a number of authors have suggested that, when we face a decision, we should choose based not just on the outcomes of the various options we might choose now, but on the joint outcome of those options and the options we anticipate we will choose when faced with future decisions – this is sometimes called *sophisticated choice* (Sahlin & Weirich, 2013; Bradley & Steele, 2014). For instance, if I use Maximin and I know that I'll face both choices in the thwartability argument for Precision given above, then I antici-pate that (i) if I refuse the first bet, and face the second using Maximin, I'll also refuse the second, and (ii) if I accept the first bet, and face the second using Maximum, I'll also accept the second. So I use sophisticated choice at the first step and accept the first bet, since accepting both dominates rejecting both.

There are a number of problems with sophisticated choice. For one thing, if we are allowed to use it to avoid the Dutch Book argument for Precision in the context of imprecise credences, we can equally use it to avoid the Dutch Book argument for Probabilism in the context of precise credences. After all, if I have credence 0.2 in $A$ and 0.2 in $\overline{A}$, and I know I'll be offered to sell a £100 bet on $A$ for £30 and then offered to sell a £100 bet on $\overline{A}$ for £30, then I can use sophisticated choice to show that I should not accept both.

Another problem is that we typically don't know which decisions we'll face in the future. We must then extend sophisticated choice so that we consider not only the decisions we know we'll face, but all the decisions we might face, perhaps weighted by our credence that we'll face them. But this threatens to make decision-making intractable.

Susanna Rinard (2015) has explored a further possibility that falls under (ii). As we noted above, your representor is the set of all and only those credence functions that agree with you in the opinions you have. Thus, if **P** is your rep-resentor, you determinately have an opinion about something iff every $p$ in **P** has that opinion as well. If there is some credence function $p$ in **P** that has that opinion and another $p'$ that doesn't, we don't say that you don't have the opin-ion, but rather that you don't determinately have that opinion. We only say that you determinately don't have an opinion if all $p$ in **P** lack it. That suggests the account of permissibility that Rinard proposes:

> **Supervaluationism** Faced with a decision problem, an available act $a$ is determinately rationally permissible for an individual with representor **P** iff, for every $p$ in **P**, $a$ is rationally permissible for an individual with credence function $p$.

This has the consequence that, for some individuals facing some decisions, there will be no option that is determinately rationally permissible – each one will be prohibited by some member of their representor.

Now, Supervaluationism does not sanction either of the Dutch Book arguments for Precision offered above. However, it also does not sanction a Dutch Book argument for Pointwise Probabilism. Indeed, provided your representor contains at least one probabilistic credence function, there can be no Dutch Book against it. For that would consist of a finite series of bets such that each is determinately permissible, but which together are guaranteed to lose money. But if they are determinately permissible, then each probabilistic credence function in the representor must render them permissible. And so the bets cannot collectively lose money, since each has positive expected value relative to the same probability function.

The upshot is that we have yet to identify a decision rule for representors that successfully navigates between the Scylla of being too permissive and thus sanctioning Dutch Book arguments for Precision, as well as other more desirable laws (such as Liberal) and the Charybdis of being too restrictive and thus sanctioning almost no Dutch Book arguments (such as Supervaluationism).

## 8 The Mathematics of the Dutch Book Arguments

A number of the arguments we presented above rely on mathematical theorems to furnish their second premises. In this section, we sketch out the proofs of those theorems.

### 8.1 Sketch Proofs of Theorems 3.2, 4.1, 7.2, and 7.3

There are three stages in each of these proofs. In the first, we see how to represent credence functions and epistemically possible worlds geometrically. In the second, we use this representation to give a geometric characterisation of the credence functions that are exploitable. In the third, we show that this geometric characterisation picks out exactly the credence functions that violate the law of credence in question.

#### 8.1.1 Representing Credence Functions and Possible Worlds Geometrically

**Representing credence functions**  Consider the credence function $c : \mathcal{F} \to [0, 1]$ defined on the finite set of propositions $\mathcal{F} = \{A_1, \ldots, A_n\}$. We write $c_i$ for $c(A_i)$ and we represent $c$ geometrically using the $n$-dimensional vector

$$c = (c(A_1), \ldots, c(A_n)) = (c_1, \ldots, c_n).$$

We abuse notation and write $c$ for both the function and vector representations.

**Representing possible worlds** Consider the epistemically possible world $w$ in $\mathcal{W}_{\mathcal{F}}^E$. Then we abuse notation and let $w$ be the indicator credence function of $w$. That is, $w : \mathcal{F} \to [0, 1]$ and:

- If classical logic governs the propositions in $\mathcal{F}$, then

$$w(A) = \begin{cases} 1 & \text{if } A \text{ is true at } w \\ 0 & \text{otherwise} \end{cases}$$

- If a non-classical logic that satisfies (T2-3) governs the propositions in $\mathcal{F}$, then

$$w(A) = \begin{cases} 1 & \text{if } A \text{ takes a designated truth value at } w \\ 0 & \text{otherwise} \end{cases}$$

- If a supervaluationist semantics governs the propositions in $\mathcal{F}$, then

$$w(A) = \begin{cases} 1 & \text{if } A \text{ is true on all precisifications at } w \\ 0 & \text{otherwise} \end{cases}$$

As above, we represent $w$ as a vector as well:

$$w = (w(A_1), \ldots, w(A_n)) = (w_1, \ldots, w_n)$$

And we abuse notation and write $w$ for the world, its indicator function, and the vector representation of that indicator function.

### 8.1.2 Characterising Exploitability Geometrically

We begin with one of the two most important mathematical results concerning Dutch Books. Both are originally due to Bruno de Finetti (1937 [1980], 1974), though each is a slight generalisation of de Finetti's version. The first characterises the credence functions that are exploitable in different ways in terms of the vector representation of credence functions given in Section 8.1.1.

**Lemma 8.1** *Suppose $\mathcal{Z}$ is a set of credence functions defined on $\mathcal{F} = \{A_1, \ldots, A_n\}$. Then:*

(i) *If $c$ is not in $\mathrm{cl}(\mathcal{Z}^+)$, then there are $S_1, \ldots, S_n$ in $\mathbb{R}$ such that $\sum_{i=1}^n z_i S_i < \sum_{i=1}^n c_i S_i$, for all $z$ in $\mathcal{Z}$.*
(ii) *If $c$ is not in $\mathrm{int}(\mathrm{cl}(\mathcal{Z}^+))$, then there are $S_1, \ldots, S_n$ in $\mathbb{R}$ such that*
   (a) *$\sum_{i=1}^n z_i S_i \leq \sum_{i=1}^n c_i S_i$, for all $z$ in $\mathcal{Z}$, and*
   (b) *$\sum_{i=1}^n z_i S_i < \sum_{i=1}^n c_i S_i$, for some $z$ in $\mathcal{Z}$.*

*Proof.* (i) This is a direct consequence of Hermann Minkowski's Separating Hyperplane Theorem (Boyd & Vandenberghe, 2004, Section 2.5). I'll sketch

Ramsey, F. P. (1926 [1931]). Truth and Probability. In R. B. Braithwaite (Ed.), *The Foundations of Mathematics and Other Logical Essays*, chap. VII (pp. 156–98). London: Kegan, Paul, Trench, Trubner & Co.

Rinard, S. (2015). A Decision Theory for Imprecise Probabilities. *Philosophers' Imprint, 15*(7), 1–16.

Rothschild, D. (2019). Lockean Beliefs, Dutch Books, and Scoring Systems. *Review of Symbolic Logic*.

Sahlin, N.-E., & Weirich, P. (2013). Unsharp Sharpness. *Theoria, 80*(1), 100–3.

Savage, L. J. (1971). Elicitation of Personal Probabilities and Expectations. *Journal of the American Statistical Association, 66*(336), 783–801.

Schervish, M. J. (1989). A General Method for Comparing Probability Assessors. *The Annals of Statistics, 17*, 1856–79.

Schick, F. (1986). Dutch Bookies and Money Pumps. *The Journal of Philosophy, 83*(2), 112–19.

Seidenfeld, T. (2004). A Contrast between Two Decision Rules for Use with (Convex) Sets of Probabilities: Gamma-Maximin versus E-Admissibility. *Synthese, 140*, 69–88.

Seidenfeld, T., Schervish, M. J., & Kadane, J. B. (1995). A Representation of Partially Ordered Preferences. *Annals of Statistics, 23*, 2168–217.

Shafer, G. (1976). *A Mathematical Theory of Evidence*. Princeton: Princeton University Press.

Skyrms, B. (1987). Coherence. In N. Rescher (Ed.), *Scientific Inquiry in Philosophical Perspective* (pp. 225–42). Pittsburgh: University of Pittsburgh Press.

Skyrms, B. (1993). A Mistake in Dynamic Coherence Arguments? *Philosophy of Science, 60*(2), 320–8.

Talbott, W. J. (1991). Two Principles of Bayesian Epistemology. *Philosophical Studies, 62*(2), 135–50.

van Fraassen, B. C. (1984). Belief and the Will. *Journal of Philosophy, 81*, 235–56.

van Fraassen, B. C. (1990). Figures in a Probability Landscape. In J. M. Dunn & A. Gupta (Eds.), *Truth or Consequences* (pp. 345–56). Dordrecht: Kluwer.

Vineberg, S. (2001). The Notion of Consistency for Partial Belief. *Philosophical Studies, 102*, 281–96.

Vineberg, S. (2016). Dutch Book Arguments. In E. N. Zalta (Ed.), *Stanford Encyclopedia of Philosophy*. Metaphysics Research Lab, Stanford University.

Walley, P. (1991). *Statistical Reasoning with Imprecise Probabilities*, vol. 42 of *Monographs on Statistics and Applied Probability*. London: Chapman and Hall.

Williams, J. R. G. (2012a). Generalized Probabilism: Dutch Books and Accuracy Domination. *Journal of Philosophical Logic*, *41*(5), 811–40.

Williams, J. R. G. (2012b). Gradational Accuracy and Non-classical Semantics. *Review of Symbolic Logic*, *5*(4), 513–37.

Williams, J. R. G. (2014). Decision-Making under Indeterminacy. *Philosophers' Imprint*, *14*(4), 1–34.

Williamson, J. (1999). Countable Additivity and Subjective Probability. *British Journal for the Philosophy of Science*, *50*, 401–16.

Wroński, L., & Godziszewski, M. T. (2017). The Stubborn Non-probabilist—'Negation Incoherence' and a New Way to Block the Dutch Book Argument. In *LORI 2017: Logic, Rationality, and Interaction*, Lecture Notes in Computer Science. Springer.

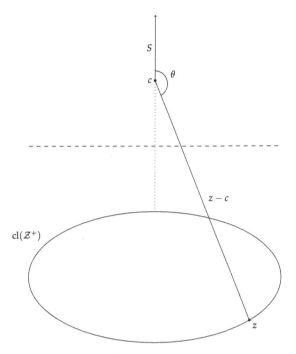

**Figure 8.1** This illustrates the Separating Hyperplane Theorem. If $c$ lies outside $\mathrm{cl}(\mathcal{Z}^+)$, there is a hyperplane that separates $c$ from $\mathrm{cl}(\mathcal{Z}^+)$. Now, take a vector $S$ that is perpendicular to that hyperplane, begins at $c$, and points away from the hyperplane. Then the angle $\theta$ between $S$ and any vector $z - c$ that runs from $c$ to a member $z$ of $\mathcal{Z}$ will be obtuse.

a geometrical proof here. Suppose $c$ is not in $\mathrm{cl}(\mathcal{Z}^+)$. Then there is a vector $S$ with $\|S\| > 0$ such that, for any $z$ in $\mathcal{Z}$, the angle $\theta$ between $S$ and $z - c$ is obtuse (see Figure 8.1 for an illustration). Thus, $\cos \theta < 0$. But we also know that $S \cdot (z - c) = \|S\| \|z - c\| \cos \theta$ and $\|S\| > 0$ and, since $c$ is not in $\mathcal{Z}^+$, $\|z - c\| > 0$. So $S \cdot (z - c) < 0$. Thus, $\sum_{i=1}^{n} z_i S_i = S \cdot z < S \cdot c = \sum_{i=1}^{n} c_i S_i$, as required.

(ii) This is a direct consequence of Minkowski's Supporting Hyperplane Theorem (Boyd & Vandenberghe, 2004, Section 2.5). The geometrical proof is similar to above (see Figure 8.2 for an illustration).

### 8.1.3 Characterising Laws of Credence Geometrically

Our second important mathematical result characterises the credences that satisfy our various laws of credence, again using the vector representation of credences given in Section 8.1.1.

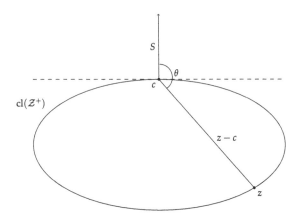

**Figure 8.2** This illustrates the Supporting Hyperplane Theorem. This time, the angle $\theta$ is either right or obtuse.

**Definition 8.2**  *Suppose $\mathcal{F}$ is a finite set of propositions and $E$ is a proposition. Then let $\mathcal{I}_{\mathcal{W}_{\mathcal{F}}^{E}}$ be the set of indicator credence functions for worlds in $\mathcal{W}_{\mathcal{F}}^{E}$.*

Some abbreviations: Full Probabilism (FP), Full Probabilism + Regularity (FPR), Full Probabilism + Strong Principal Principle (FPP), Full Probabilism + Strong Principal Principle$^{+}$ (FPPP$^{+}$), Full Generalised Probabilism (FGP), Supervaluational Dempster-Shaferism (SDS).

First, we note that, if a credence function $c$ defined on infinite set of propositions $\mathcal{F}$ violates one of the laws in question, there is a finite subset $\mathcal{F}' \subseteq \mathcal{F}$ such that the credences that $c$ assigns to the propositions in $\mathcal{F}'$ already violate the law.

**Lemma 8.3**  *Suppose $c : \mathcal{F} \to [0, 1]$. Then*

$$
c \text{ satisfies} \left\{ \begin{array}{l} FP \\ FPR \\ FPPP \\ FGP \end{array} \right\} \text{ iff, for all finite } \mathcal{F}' \subseteq \mathcal{F}, \ c|_{\mathcal{F}'} \text{ satisfies} \left\{ \begin{array}{l} FP \\ FPR \\ FPPP \\ FGP \end{array} \right\}
$$

In the light of this, and since SDS is stated in terms of finite algebras in the first place, we will focus only on credence functions on finite sets of propositions in what follows.

Our next result characterises the credence functions defined on finite sets of propositions that satisfy our various laws of credence.

**Lemma 8.4**   *Suppose $c : \mathcal{F} \to [0, 1]$ and $\mathcal{F}$ is finite. Then:*

- *If $\mathcal{F}$ is governed by classical logic, then:*
    - (i) *$c$ satisfies FP iff $c$ is in $(\mathcal{I}_{\mathcal{W}^E_\mathcal{F}})^+$ (de Finetti, 1974; Predd et al., 2009).*
    - (ii) *$c$ satisfies FPR iff $c$ is in $\mathrm{int}((\mathcal{I}_{\mathcal{W}^E_\mathcal{F}})^+)$.*
    - (iii) *$c$ satisfies FPPP iff $c$ is in $\mathcal{C}^+$, if $\mathcal{C}$ is finite (Pettigrew, 2012, 2013).*
    - (iv) *$c$ satisfies FPPP$^+$ iff $c$ is in $\mathrm{cl}(\mathcal{C}^+)$.*
- *If $\mathcal{F}$ is governed by a non-classical logic that satisfies (T2-3), then:*
    - (v) *$c$ satisfies FGP iff $c$ is in $(\mathcal{I}_{\mathcal{W}^E_\mathcal{F}})^+$ (Paris, 2001; Bradley, 2017).*
- *If $\mathcal{F}$ is governed by supervaluationist semantics, then:*
    - (vi) *$c$ satisfies SDS iff $c$ is in $(\mathcal{I}_{\mathcal{W}^E_\mathcal{F}})^+$ (Jaffray, 1989; Bradley, 2017).*

### 8.1.4 Building the Proofs

We now have the ingredients to prove Theorems 3.2, 4.1, 7.2, and 7.3. Here, I will sketch the proof of Theorem 3.2(DB2FP). The proofs of Theorems 3.2(DB2FPR), 3.2(DB2FPPP), 3.2(DB2FPPP$^+$), 7.2, and 7.3 are similar. I will then sketch the proof of Theorem 4.1.

*Sketch proof for Theorem 3.2(DB2FP)*
First, the left-to-right direction. Suppose $c$ violates Full Probabilism. Then, by Lemma 8.4(i), $c$ is not in $(\mathcal{I}_{\mathcal{W}^E_\mathcal{F}})^+$. Since $\mathcal{I}_{\mathcal{W}^E_\mathcal{F}}$ is finite, $(\mathcal{I}_{\mathcal{W}^E_\mathcal{F}})^+$ is already closed, and so $(\mathcal{I}_{\mathcal{W}^E_\mathcal{F}})^+ = \mathrm{cl}(\mathcal{I}^+_{\mathcal{W}^E_\mathcal{F}})$. Thus, letting $\mathcal{Z} = \mathcal{I}_{\mathcal{W}^E_\mathcal{F}}$ in Lemma 8.1(i), there are $S_1, \ldots, S_n$ in $\mathbb{R}$ such that $\sum_{i=1}^n w_i S_i < \sum_{i=1}^n c_i S_i$, for all $w$ in $\mathcal{W}^E_\mathcal{F}$. It follows that there are $x_1, \ldots, x_n$ in $\mathbb{R}$ such that:

(i) $x_i < c_i S_i$ for $1 \leq i \leq n$, and
(ii) $\sum_{i=1}^n w_i S_i < \sum_{i=1}^n x_i$, for all $w$ in $\mathcal{W}^E_\mathcal{F}$.

Now, by (i) and (DB1), $c$ requires that you pay £$x_i$ for a £$S_i$ bet on $A_i$. So you will pay out £$\sum_{i=1}^n x_i$ for the book of these bets. But a £$S_i$ bet on $A_i$ pays out $w_i S_i$ at world $w$. So the total net gain of the bets taken together at world $w$ will be £$\left(\sum_{i=1}^n w_i S_i - \sum_{i=1}^n x_i\right) <$ £0. So, by (ii), $c$ is finitely strongly fully exploitable.

Second, the right-to-left direction. Suppose $c$ satisfies Full Probabilism. Then, by Lemma 8.4(i), $c$ is in $(\mathcal{I}_{\mathcal{W}^E_\mathcal{F}})^+$. So $c = \sum_{w \in \mathcal{W}^E_\mathcal{F}} \lambda_w w$. Then suppose for the sake of proving a contradiction that there are $S_1, \ldots, S_n$ and $x_1, \ldots, x_n$ in $\mathbb{R}$ such that $x_i \leq c_i S_i$, for all $1 \leq i \leq n$, but $\sum_{i=1}^n w_i S_i < \sum_{i=1}^n x_i \leq \sum_{i=1}^n c_i S_i$, for all $w$ in $\mathcal{W}^E_\mathcal{F}$. So

$$0 > \sum_{w \in \mathcal{W}^E_\mathcal{F}} \lambda_w \left( \sum_{i=1}^n w_i S_i - \sum_{i=1}^n c_i S_i \right)$$

$$= \sum_{i=1}^{n} \left( \sum_{w \in \mathcal{W}_{\mathcal{F}}^{E}} \lambda_w w_i \right) S_i - \sum_{i=1}^{n} c_i S_i$$

$$= \sum_{i=1}^{n} c_i S_i - \sum_{i=1}^{n} c_i S_i = 0$$

which gives a contradiction. So $c$ is not finitely weakly fully exploitable. $\square$

- To prove Theorem 3.2(DB2FPR), replace Lemma 8.4(i) with Lemma 8.4(ii), and Lemma 8.1(i) with Lemma 8.1(ii).
- To prove Theorem 3.2(DB2FPPP), let $\mathcal{Z}$ be the set of epistemically possible chance functions in Lemma 8.1(i), replace Lemma 8.4(i) with Lemma 8.4(iii), and note that, while $£\sum_{i=1}^{n} w_i S_i$ gives the total payout of the set of bets at $w$, $£\sum_{i=1}^{n} ch_i^k S_i$ gives the total expected payout of the set of bets by the lights of the chance function $ch^k$.
- To prove Theorem 3.2(DB2FPPP$^+$), let $\mathcal{Z}$ be the set of epistemically possible chance functions in Lemma 8.1(i), replace Lemma 8.4(i) with Lemma 8.4(iv), and proceed as above.
- To prove Theorem 7.2, replace Lemma 8.4(i) with Lemma 8.4(v).
- To prove Theorem 7.3, replace Lemma 8.4(i) with Lemma 8.4(vi).

*Sketch proof of Theorem 4.1*
First, we state and prove the following lemma characterising the super-conditionalising rules for a prior:

## Lemma 8.5

(i) *R is a weak super-conditionalising rule for c iff, for each $E_i$ in $\mathcal{E}$ and $c'$ in $C_i$, there is $0 \leq \lambda_{c'}^i \leq 1$ with $\sum_{E_i \in \mathcal{E}} \sum_{c' \in C_i} \lambda_{c'}^i = 1$ such that*
  (a) *for all $E_i$ in $\mathcal{E}$ and $c'$ in $C_i$, if $\lambda_{c'}^i > 0$, then $c'(E_i) = 1$, and*
  (b) $c(-) = \sum_{E_i \in \mathcal{E}} \sum_{c' \in C_i} \lambda_{c'}^i c'(-)$.
(ii) *R is a strong super-conditionalising rule for c iff, for each $E_i$ in $\mathcal{E}$ and $c'$ in $C_i$, there is $0 < \lambda_{c'}^i < 1$ with $\sum_{E_i \in \mathcal{E}} \sum_{c' \in C_i} \lambda_{c'}^i = 1$ such that*
  (a) *for all $E_i$ in $\mathcal{E}$ and $c'$ in $C_i$, $c'(E_i) = 1$; and*
  (b) $c(-) = \sum_{E_i \in \mathcal{E}} \sum_{c' \in C_i} \lambda_{c'}^i c'(-)$.

*Proof.* Let's take (i) first. We'll begin with the left-to-right direction.
Suppose $R$ is a weak super-conditionalising rule for $c$. Then for each $E_i$ in $\mathcal{E}$ and $c'$ in $C_i$, let $\lambda_{c'}^i = c^\dagger(R_{c'}^i)$. Then, if $\lambda_{c'}^i > 0$, then $c^\dagger(R_{c'}^i) > 0$, and so $c'(E_i) = c^\dagger(E_i|R_{c'}^i)$. But $R_{c'}^i$ says that you received evidence $E_i$ and responded by adopting credence function $c'$. So $R_{c'}^i$ entails $E_i$, and thus $c^\dagger(E_i|R_{c'}^i) = 1$. So $c'(E_i) = 1$. That gives (a).

Cambridge Elements $\equiv$

# Decision Theory and Philosophy

Elements in the Series

*Dutch Book Arguments*
Richard Pettigrew

A full series listing is available at: www.cambridge.org/EDTP

Cambridge Elements ☰

# Decision Theory and Philosophy

## Martin Peterson

*Texas A&M University*

Martin Peterson is Professor of Philosophy and Sue and Harry E. Bovay Professor of the History and Ethics of Professional Engineering at Texas A&M University. He is the author of four books and one edited collection, as well as many articles on decision theory, ethics and philosophy of science.

## About the Series

This Cambridge Elements series offers an extensive overview of decision theory in its many and varied forms. Distinguished authors provide an up-to-date summary of the results of current research in their fields and give their own take on what they believe are the most significant debates influencing research, drawing original conclusions.

Now, for each $E_i$ in $\mathcal{E}$ and $c'$ in $C_i$, and for each possible world $w$, we have $c^\dagger(w \,\&\, R^i_{c'}) = c^\dagger(R^i_{c'})c'(w)$. Thus:

$c(w) = c^\dagger(w)$  since $c^\dagger$ extends $c$

$$= \sum_{E_i \in \mathcal{E}} \sum_{c' \in C_i} c^\dagger(w \,\&\, R^i_{c'}) \text{ by Finite Additivity of } c^*$$

$$= \sum_{E_i \in \mathcal{E}} \sum_{c' \in C_i} c^\dagger(R^i_{c'})c'(w) \text{ as noted above}$$

$$= \sum_{E_i \in \mathcal{E}} \sum_{c' \in C_i} \lambda^i_{c'} c'(w)$$

as required. This gives (b).

Second, we take the right-to-left direction of (i). Suppose (a) and (b) hold. Then there is, for each $E_i$ in $\mathcal{E}$ and $c'$ in $C_i$, $0 \le \lambda^i_{c'} \le 1$ with $\sum_{E_i \in \mathcal{E}} \sum_{c' \in C_i} \lambda^i_{c'} = 1$ such that

$$c(-) = \sum_{E_i \in \mathcal{E}} \sum_{c' \in C_i} \lambda^i_{c'} c'(-)$$

So, given a possible world $w$, $E_i$ in $\mathcal{E}$, and $c'$ in $C_i$, let

$$c^\dagger(w \,\&\, R^i_{c'}) = \lambda^i_{c'} c'(w)$$

Then

- For any possible world $w$,

$$c^\dagger(w) = \sum_{E_i \in \mathcal{E}} \sum_{c' \in C_i} c^\dagger(w \,\&\, R^i_{c'}) = \sum_{E_i \in \mathcal{E}} \sum_{c' \in C_i} \lambda^i_{c'} c'(w) = c(w)$$

  So $c^\dagger$ is an extension of $c$.

- For any possible world $w$, $E_i$ in $\mathcal{E}$, and $c'$ in $C_i$, if $c^\dagger(R^i_{c'}) > 0$, then

$$c^\dagger(w|R^i_{c'}) = \frac{c^\dagger(w \,\&\, R^i_{c'})}{c^\dagger(R^i_{c'})} = \frac{\lambda^i_{c'} c'(w)}{\sum_{w' \in W} \lambda^i_{c'} c'(w')} = \frac{\lambda^i_{c'} c'(w)}{\lambda^i_{c'} \sum_{w' \in W} c'(w')} = c'(w)$$

  and thus $c'(E_i) = c^\dagger(E_i|R^i_{c'}) = 1$.

Thus, $R$ is a weak super-conditionalising rule for $c$. This establishes Lemma 8.5(i). The proof of Lemma 8.5(ii) proceeds similarly. $\qquad\square$

Next, we move to the proof of Theorem 4.1.

First, Theorem 4.1(I). Suppose $R$ is not a weak or a strong super-conditionalising rule for $c$. Then, either

(a) $c'(E_i) < 1$ for some $E_i$ in $\mathcal{E}$ and $c'$ in $C_i$; or
(b) $c$ is not in $\{c' : E_j \in \mathcal{E} \,\&\, c' \in C_j\}^+$.

Let's take these in turn.

First, (a). Suppose $c'(E_i) < p < 1$ for some $E_i$ in $\mathcal{E}$ and $c'$ in $C_i$. Then offer no bets at the earlier time, and, if you receive evidence $E_i$ and respond with credence function $c'$, you are required to sell a £1 bet on $E_i$ for £$p$, which will lose you £$(1 - p) > 0$ at all worlds in $E_i$. This gives a weak Dutch Strategy against you.

Second, (b). Suppose $c$ is not in $\{c' : E_j \in \mathcal{E} \,\&\, c' \in C_j\}^+$. Then, by Lemma 8.1(i), there are $S_1, \dots, S_n$ in $\mathbb{R}$ such that $\sum_{i=1}^{n} c'_i S_i < \sum_{i=1}^{n} c_i S_i$, for all $E_j$ in $\mathcal{E}$ and $c'$ in $C_j$. It follows that there are $x_1, \dots, x_n$ and $y_1, \dots, y_n$ in **R** such that, for each $E_j$ in $\mathcal{E}$ and $c'$ in $C_j$,

(i) $x_i < c_i S_i$ for all $1 \leq i \leq n$,
(ii) $c'_i S_i < y_i$ for all $1 \leq i \leq n$, and
(iii) $\sum_{i=1}^{n} y_i < \sum_{i=1}^{n} x_i$.

Now, by (i), (ii), and (DB1), $c$ requires you to buy a £$S_i$ bet on $A_i$ for £$x_i$, and $c'$ requires you to sell a £$S_i$ bet on $A_i$ for £$y_i$. Thus, your total net gain will be £$\left(\sum_{i=1}^{n} y_i - \sum_{i=1}^{n} x_i\right) < 0$ at every world. So $c$ and $R$ are together vulnerable to a strong Dutch Strategy.

Second, Theorem 4.1(II). Suppose $R$ is a strong super-conditionalising rule for $c$. So $c$ defined on $\mathcal{F}$ can be extended to $c^\dagger$ defined on $\mathcal{F}^\dagger$ such that $c^\dagger(R^j_{c'}) > 0$ and $c'(A) = c^\dagger(A|R^j_{c'})$. Then suppose for a contradiction that

(i) there are $S_1, \dots, S_n$ and $x_1, \dots, x_n$ in $\mathbb{R}$ and
(ii) for each $E_j$ in $\mathcal{E}$ and $c'$ in $C_j$, there are $S'_1, \dots, S'_n$ and $y'_1, \dots, y'_n$ in $\mathbb{R}$,

such that

(iii) $x_i \leq c_i S_i$, for all $1 \leq i \leq n$,
(iv) $y'_i \leq c'_i S'_i$, for all $1 \leq i \leq n$, and for each $E_j$ in $\mathcal{E}$ and $c'$ in $C_j$, and
(v) for each $E_j$ in $\mathcal{E}$ and $c'$ in $C_j$ and $w$ in $E_j$,

$$\left(\sum_{i=1}^{n} w_i S_i - \sum_{i=1}^{n} x_i\right) + \left(\sum_{i=1}^{n} w_i S'_i - \sum_{i=1}^{n} y'_i\right) < 0$$

Then, for each $E_j$ in $\mathcal{E}$ and $c'$ in $C_j$ and $w$ in $E_j$,

$$\left(\sum_{i=1}^{n} w_i S_i - \sum_{i=1}^{n} c_i S_i\right) + \left(\sum_{i=1}^{n} w_i S'_i - \sum_{i=1}^{n} c'_i S'_i\right) < 0$$

So,

$$
\begin{aligned}
0 > &\sum_{E_j \in \mathcal{E}} \sum_{c' \in C_j} \sum_{w \in E_j} c^\dagger(w \,\&\, R_{c'}^j) \left( \left( \sum_{i=1}^n w_i S_i - \sum_{i=1}^n c_i S_i \right) + \left( \sum_{i=1}^n w_i S_i' - \sum_{i=1}^n c_i' S_i' \right) \right) \\
= &\sum_{E_j \in \mathcal{E}} \sum_{c' \in C_j} \sum_{w \in E_j} c^\dagger(w | R_{c'}^j) c^\dagger(R_{c'}^j) \left( \left( \sum_{i=1}^n w_i S_i - \sum_{i=1}^n c_i S_i \right) + \left( \sum_{i=1}^n w_i S_i' - \sum_{i=1}^n c_i' S_i' \right) \right) \\
= &\sum_{E_j \in \mathcal{E}} \sum_{c' \in C_j} \sum_{w \in E_j} c'(w) c^\dagger(R_{c'}^j) \left( \left( \sum_{i=1}^n w_i S_i - \sum_{i=1}^n c_i S_i \right) + \left( \sum_{i=1}^n w_i S_i' - \sum_{i=1}^n c_i' S_i' \right) \right) \\
= &\sum_{E_j \in \mathcal{E}} \sum_{c' \in C_j} c^\dagger(R_{c'}^j) \sum_{w \in E_j} c'(w) \left( \left( \sum_{i=1}^n w_i S_i - \sum_{i=1}^n c_i S_i \right) + \left( \sum_{i=1}^n w_i S_i' - \sum_{i=1}^n c_i' S_i' \right) \right) \\
= &\sum_{E_j \in \mathcal{E}} \sum_{c' \in C_j} c^\dagger(R_{c'}^j) \left( \left( \sum_{i=1}^n \left( \sum_{w \in E_j} c'(w) w_i \right) S_i - \sum_{i=1}^n c_i S_i \right) + \right. \\
&\left. \left( \sum_{i=1}^n \left( \sum_{w \in E_j} c'(w) w_i \right) S_i' - \sum_{i=1}^n c_i' S_i' \right) \right) \\
= &\sum_{E_j \in \mathcal{E}} \sum_{c' \in C_j} c^\dagger(R_{c'}^j) \left( \left( \sum_{i=1}^n c_i' S_i - \sum_{i=1}^n c_i S_i \right) + \left( \sum_{i=1}^n c_i' S_i' - \sum_{i=1}^n c_i' S_i' \right) \right) \\
= &\sum_{E_j \in \mathcal{E}} \sum_{c' \in C_j} c^\dagger(R_{c'}^j) \left( \sum_{i=1}^n c_i' S_i - \sum_{i=1}^n c_i S_i \right) \\
= &\sum_{i=1}^n \left( \sum_{E_j \in \mathcal{E}} \sum_{c' \in C_j} c^\dagger(R_{c'}^j) c_i' \right) S_i - \sum_{i=1}^n c_i S_i \\
= &\sum_{i=1}^n \left( \sum_{E_j \in \mathcal{E}} \sum_{c' \in C_j} c^\dagger(R_{c'}^j) c^\dagger(A_i | R_{c'}^j) \right) S_i - \sum_{i=1}^n c_i S_i \\
= &\sum_{i=1}^n \left( \sum_{E_j \in \mathcal{E}} \sum_{c' \in C_j} c^\dagger(A_i \,\&\, R_{c'}^j) \right) S_i - \sum_{i=1}^n c_i S_i \\
= &\sum_{i=1}^n c^\dagger(A_i) S_i - \sum_{i=1}^n c_i S_i \\
= &\sum_{i=1}^n c_i S_i - \sum_{i=1}^n c_i S_i \\
= &\, 0
\end{aligned}
$$

which gives a contradiction. So $c$ and $R$ are not together vulnerable to a weak Dutch Strategy. The proof of Theorem 4.1(III) proceeds similarly. $\qquad\square$

## 8.2 Sketch Proof of Theorem 7.1

The following facts allow us to derive Theorem 7.1 from Theorem 4.1:

First, let

$$c^*(w) = \frac{\frac{c(w)}{N(w)}}{\sum_{w' \in \mathcal{W}} \frac{c(w')}{N(w')}} \quad \text{and} \quad c_i^*(w) = \frac{\frac{c_i(w)}{N_i(w)}}{\sum_{w' \in \mathcal{W}} \frac{c_i(w')}{N_i(w')}}$$

Then:

$$\left\{\begin{array}{l} \text{there are } a^*, a_1^*, \ldots, a_n^* \text{ such that} \\ \mathrm{EU}^{\mathrm{CDT}}_{c^*,u}(a^*) > 0 \\ \mathrm{EU}^{\mathrm{CDT}}_{c_i^*,u}(a_i^*) > 0 \\ u(a^*, w) + u(a_i^*, w) < 0, \\ \quad \text{for } 1 \leq i \leq n \end{array}\right\} \Leftrightarrow \left\{\begin{array}{l} \text{There are } a, a_1, \ldots a_n \text{ such that} \\ \mathrm{EU}^{\mathrm{CDT}}_{c,u}(a) > 0 \\ \mathrm{EU}^{\mathrm{CDT}}_{c_i,u}(a_i) > 0 \\ N(w)u(a, w) + N_i(w)u(a_i, w) < 0, \\ \quad \text{for } 1 \leq i \leq n \end{array}\right\}$$

And:

$$\left\{\begin{array}{l} \text{there are } a^*, a_1^*, \ldots, a_n^* \text{ such that} \\ \mathrm{EU}^{\mathrm{CDT}}_{c,u}(a^*) > 0 \\ \mathrm{EU}^{\mathrm{CDT}}_{c_i,u}(a_i^*) > 0 \\ u(a^*, w) + u(a_i^*, w) < 0, \\ \quad \text{for } 1 \leq i \leq n \end{array}\right\} \Leftrightarrow \left\{\begin{array}{l} \text{there are } a, a_1, \ldots a_n \text{ such that} \\ \mathrm{EU}^{\mathrm{EDT}}_{c,u}(a) > 0 \\ \mathrm{EU}^{\mathrm{EDT}}_{c_i,u}(a_i) > 0 \\ N(w)u(a, w) + N_i(w)u(a_i, w) < 0, \\ \quad \text{for } 1 \leq i \leq n \end{array}\right\}$$

## 8.3 Sketch Proof of Theorem 5.1

(I) Suppose $c$ violates Bounded Probabilism. There are three possibilities:

(i) $c(\bot) > 0$. Then, by (MEU), $c$ rationally requires you to pay $£\frac{c(\bot)}{2(c(\bot)+c(\top))}$ for a £1 bet on $\bot$.

(ii) $c(\bot) = c(\top) = 0$. Then, by (MEU), $c$ rationally permits you to pay £1 for a £1 bet on $\bot$.

(iii) $c(X \vee Y) \neq c(X) + c(Y)$ for mutually exclusive $X, Y$. Suppose $c(X \vee Y) < c(X) + c(Y)$. So

$$c(X \vee Y) + c(\overline{X \vee Y}) = x < y = c(X) + c(Y) + c(\overline{X \vee Y}).$$

Thus, we define two options $a$ and $b$ as follows. First, pick five numbers $x < z_1 < z_2 < z_3 < z_4 < z_5 < y$. Then

- $a$ has two outcomes $o_1$ and $o_2$:

$$\begin{array}{lclcl} (a \Rightarrow o_1) & = & X \vee Y & \text{and} & u(o_1) & = & z_5 \\ (a \Rightarrow o_2) & = & \overline{X \vee Y} & \text{and} & u(o_2) & = & z_4 \end{array}$$

- $b$ has two outcomes $o'_1, o'_2, o'_3$:

$$
\begin{aligned}
(b \Rightarrow o'_1) &= X & \text{and} && u(o'_1) &= z_1 \\
(b \Rightarrow o'_2) &= Y & \text{and} && u(o'_2) &= z_2 \\
(b \Rightarrow o'_3) &= \overline{X \vee Y} & \text{and} && u(o'_3) &= z_3
\end{aligned}
$$

Then note:

(i) $a$ dominates $b$

After all, $u(o'_1), u(o'_2), u(o'_3) < u(o_1), u(o_2)$.

(ii) $EU_{c,u}(b) > EU_{c,u}(a)$.

After all,

$$
\begin{aligned}
EU_{c,u}(a) &= c(a \Rightarrow o_1)u(o_1) + c(a \Rightarrow o_2)u(o_2) \\
&= c(X \vee Y)z_5 + c(\overline{X \vee Y})z_4 \\
&< (c(X \vee Y) + c(\overline{X \vee Y}))y \\
&= xy
\end{aligned}
$$

$$
\begin{aligned}
EU_{c,u}(b) &= c(b \Rightarrow o_1)u(o_1) + c(b \Rightarrow o_2)u(o_2) + c(b \Rightarrow o_3)u(o_3) \\
&= c(X)z_1 + c(Y)z_2 + c(\overline{X \vee Y})z_3 \\
&> (c(X) + c(Y) + c(\overline{X \vee Y}))x \\
&= xy
\end{aligned}
$$

And similarly if $c(X \vee Y) > c(X) + c(Y)$.

Second, (II). Suppose $c$ satisfies Bounded Probabilism. Then, there is $0 < M \leq 1$ and a probabilistic credence function $p$ such that $c(-) = M \times p(-)$. And, for any option $a$,

$$
EU_{c,u}(a) = EU_{Mp,u}(a) = M \times EU_{p,u}(a)
$$

Thus, $EU_{c,u}(a) > EU_{c,u}(b)$ iff $EU_{p,u}(a) > EU_{p,u}(b)$. Since $p$ neither requires or permits you to choose a dominated option, neither does $c$. □

# References

Allais, M. (1953). Le comportement de l'homme rationnel devant le risque: critique des postulats et axiomes de l'école Américaine. *Econometrica*, *21*(4), 503–46.

Armendt, B. (1993). Dutch Books, Additivity, and Utility Theory. *Philosophical Topics*, *21*(1).

Arntzenius, F. (2002). Reflections on Sleeping Beauty. *Analysis*, *62*(2), 53–62.

Arntzenius, F. (2003). Some Problems for Conditionalization and Reflection. *Journal of Philosophy*, *100*(356–70).

Arntzenius, F., Elga, A., & Hawthorne, J. (2004). Bayesianism, Infinite Decisions, and Binding. *Mind*, *113*, 251–83.

Berger, J. O. (1985). *Statistical Decision Theory and Bayesian Analysis*. New York: Springer.

Bernoulli, D. (1738 [1954]). Exposition of a New Theory on the Measurement of Risk. *Econometrica*, *22*(1), 23–36.

Boyd, S., & Vandenberghe, L. (2004). *Convex Optimization*. Cambridge, UK: Cambridge University Press.

Bradley, S. (2016). Imprecise Probabilities. In E. N. Zalta (Ed.), *Stanford Encyclopedia of Philosophy*. Metaphysics Research Lab, Stanford University.

Bradley, S. (2017). Nonclassical Probability and Convex Hulls. *Erkenntnis*, *82*(1), 87–101.

Bradley, S., & Steele, K. (2014). Should Subjective Probabilities Be Sharp? *Episteme*, *11*(3), 277–89.

Briggs, R. A. (2009). Distorted Reflection. *Philosophical Review*, *118*(1), 59–85.

Briggs, R. A. (2010). Putting a Value on Beauty. In T. S. Gendler & J. Hawthorne (Eds.), *Oxford Studies in Epistemology*, vol. 3 (pp. 3–34). Oxford University Press.

Briggs, R. A., & Pettigrew, R. (2018). An Accuracy-Dominance Argument for Conditionalization. *Noûs*.

Buchak, L. (2013). *Risk and Rationality*. Oxford, UK: Oxford University Press.

Christensen, D. (1991). Clever Bookies and Coherent Beliefs. *Philosophical Review*, *100*(2), 229–47.

Christensen, D. (1996). Dutch-Book Arguments Depragmatized: Epistemic Consistency for Partial Believers. *The Journal of Philosophy*, *93*(9), 450–79.

Davidson, D., McKinsey, J. C. C., & Suppes, P. (1955). Outlines of a Formal Theory of Value, I. *Philosophy of Science*, *22*(2), 140–60.

de Finetti, B. (1937 [1980]). Foresight: Its Logical Laws, Its Subjective Sources. In H. E. Kyburg & H. E. K. Smokler (Eds.), *Studies in Subjective Probability*. Huntington, NY: Robert E. Kreiger Publishing Co.

de Finetti, B. (1972). *Probability, Induction, and Statistics*. London: John Wiley & Sons.

de Finetti, B. (1974). *Theory of Probability*, vol. I. New York: John Wiley & Sons.

Dempster, A. P. (1968). A Generalization of Bayesian Inference. *Journal of the Royal Statistical Society Series B (Methodological)*, *30*, 205–47.

Easwaran, K. (2016). Dr Truthlove, Or: How I Learned to Stop Worrying and Love Bayesian Probabilities. *Noûs*, *50*(4), 816–53.

Easwaran, K., & Fitelson, B. (2015). Accuracy, Coherence, and Evidence. In T. S. Gendler & J. Hawthorne (Eds.), *Oxford Studies in Epistemology*, vol. 5, (pp. 61–96). Oxford University Press.

Elga, A. (2000). Self-Locating Belief and the Sleeping Beauty Problem. *Analysis*, *60*(2), 143–7.

Elga, A. (2010). Subjective Probabilities Should Be Sharp. *Philosophers' Imprint*, *10*(5), 1–11.

Fishburn, P. C. (1986). The Axioms of Subjective Probability. *Statistical Science*, *1*(3), 335–58.

Gerla, B. (2000). MV-Algebras, Multiple Bets and Subjective States. *International Journal of Approximate Reasoning*, *25*, 1–13.

Gilboa, I., & Schmeidler, D. (1989). Maxmin Expected Utility with Non-Unique Prior. *Journal of Mathematical Economics*, *18*, 141–53.

Hájek, A. (2008). Dutch Book Arguments. In P. Anand, P. Pattanaik, & C. Puppe (Eds.), *The Oxford Handbook of Rational and Social Choice* (pp. 173–95). Oxford: Oxford University Press.

Hedden, B. (2013). Incoherence without Exploitability. *Noûs*, *47*(3), 482–495.

Howson, C., & Urbach, P. (1989). *Scientific Reasoning: The Bayesian Approach*. La Salle, IL: Open Court.

Jaffray, J.-Y. (1989). Coherent Bets under Partially Resolving Uncertainty and Belief Functions. *Theory and Decision*, *26*, 99–105.

Jeffrey, R. C. (1983). *The Logic of Decision*. 2nd ed. Chicago and London: University of Chicago Press.

Joyce, J. M. (1999). *The Foundations of Causal Decision Theory*. Cambridge Studies in Probability, Induction, and Decision Theory. Cambridge: Cambridge University Press.

Joyce, J. M. (2010). A Defense of Imprecise Credences in Inference and Decision Making. *Philosophical Perspectives*, *24*, 281–322.

Levi, I. (1974). On Indeterminate Probabilities. *Journal of Philosophy*, *71*, 391–418.

Levinstein, B. A. (2017). A Pragmatist's Guide to Epistemic Utility. *Philosophy of Science*, *84*(4), 613–38.

Lewis, D. (1980). A Subjectivist's Guide to Objective Chance. In R. C. Jeffrey (Ed.), *Studies in Inductive Logic and Probability*, vol. II. Berkeley: University of California Press.

Lewis, D. (1981). Causal Decision Theory. *Australasian Journal of Philosophy*, *59*, 5–30.

Lewis, D. (1999). Why Conditionalize? In *Papers in Metaphysics and Epistemology* (pp. 403–7). Cambridge, UK: Cambridge University Press.

Lewis, D. (2001). Sleeping Beauty: Reply to Elga. *Analysis*, *61*(3), 171–6.

MacFarlane, J., & Kolodny, N. (2010). Ifs and Oughts. *Journal of Philosophy*, *107*, 115–43.

Mahtani, A. (2012). Diachronic Dutch Book Arguments. *Philosophical Review*, *121*(3), 443–50.

Mahtani, A. (2015). Dutch Books, Coherence, and Logical Consistency. *Noûs*, *49*(3), 522–37.

McGee, V. (1999). An Airtight Dutch Book. *Analysis*, *59*(4), 257–65.

Moss, S. (2015). Credal Dilemmas. *Noûs*, *49*(4), 665–83.

Moss, S. (2018). *Probabilistic Knowledge*. Oxford, UK: Oxford University Press.

Parfit, D. (ms). What We Together Do.

Paris, J. B. (2001). A Note on the Dutch Book Method. In *Proceedings of the 2nd International Symposium on Imprecise Probabilities and Their Applications, ISIPTA, Ithaca, NY* (pp. 301–6). Oxford, UK: Shaker.

Pettigrew, R. (2012). Accuracy, Chance, and the Principal Principle. *Philosophical Review*, *121*(2), 241–75.

Pettigrew, R. (2013). A New Epistemic Utility Argument for the Principal Principle. *Episteme*, *10*(1), 19–35.

Pettigrew, R. (2019a). On the Expected Utility Objection to the Dutch Book Argument for Probabilism. *Noûs* (DOI: 10.1111/nous.12286).

Pettigrew, R. (2019b). Is Conditionalization, and Why Should We Do It? *Philosophical Studies* (DOI: 10.1007/s11098-019-01377-y).

Piccione, M. & Rubinstein, A. (1997). On the Interpretation of Decision Problems with Imperfect Recall. *Games and Economical Behavior*, *20*, 3–24.

Predd, J., Seiringer, R., Lieb, E. H., Osherson, D., Poor, V., & Kulkarni, S. (2009). Probabilistic Coherence and Proper Scoring Rules. *IEEE Transactions of Information Theory*, *55*(10), 4786–92.

# Acknowledgements

While there's just one author credited on the dust jacket, this Element is really a collective effort. The way I present the material in the Element owes a debt to the papers and surveys I've read over the years, the talks I've attended, and the conversations I've had with colleagues at the whiteboard, where they've patiently explained an argument or sketched out the proof of a theorem. I have found the community of formal epistemologists to be splendidly collegial, and I wish to acknowledge all the ways they put me in a position to write this Element.

As well as this general debt to all those who have contributed to my understanding, I am particularly grateful to those who gave specific and detailed feedback on the manuscript. Seamus Bradley sent generous comments on the whole manuscript from a reading group he held with Jack Woods, Ed Elliott, and Robbie Williams. Samir Okasha also gave insightful comments on a full draft, as did Martin Peterson and two generous referees for Cambridge University Press. Brian Hedden, Leszek Wroński, and Michał Tomasz Godziszewski indulged me in a lengthy exchange about Section 5. And Catrin Campbell-Moore and Jason Konek helped me greatly as I tried to corral the material for Section 7.